AF440656

Quantitative Aspects of Chemical Pharmacology

CHEMICAL IDEAS IN DRUG ACTION
WITH NUMERICAL EXAMPLES

R.B. BARLOW, MA, BSc., D Phil.

Lecturer in Chemical Pharmacology,
University of Bristol

CROOM HELM LONDON

British Library Cataloguing in Publication Data

Barlow, R.B.
    Quantitative aspects of chemical pharmacology.
    1. Structure-activity relationship
    (Pharmacology)
    I. Title
    615'.7   QP906.S75

0-85664-892-2
0-7099-0300-6(pb)

Reproduced from copy supplied
printed and bound in Great Britain
by Billing and Sons Limited
Guildford, London, Oxford, Worcester

Contents

References are numbered and listed at the end of each chapter. The Tables for chapter 1 contain statistical information and are collected together on pages 36—8. The other Tables and all the Figures are inserted into the text where appropriate.

Acknowledgements

I wish to thank the following for permission to include material which is copyright:

Dr R. C. Campbell and the Cambridge University Press for Tables 1.1, 1.2 and 1.3, which are shortened versions of Tables A.12, A.15 and 16, and A.3 in *Statistics for Biologists*; the Editors of the *Journal of Pharmacy and Pharmacology* for Table 1.4, which is taken from the paper by C. I. Bliss (1938), *Quart. J. Pharmac.*, no. 11, p. 195; Prof. B. Pullman and Academic Press for Figure 3.16 which is taken from the paper by J. L. Coubeils and B. Pullman (1972) *Mol. Pharmacol.*, no. 8, p. 281; the Editorial Board of the *British Journal of Pharmacology* for Figure 5.4 which is taken from the paper by F. B. Abramson, R. B. Barlow, M. G. Mustafa and R. P. Stephenson (1969) *Br. J. Pharmac.*, no. 37, p. 223.

I am most grateful to Dr P. S. Brown and Dr F. Roberts for valuable comments on Chapter 1 and to Dr Brown, Dr P. V. Taberner, Dr N. J. M. Birdsall and Prof. B. L. Ginsborg for material used in the examples. I thank Prof. J. F. Mitchell for his encouragement.

Finally I wish to thank Mrs Pat Berman for her expert typing and Mrs Anne Duncan for the illustrations.

I should be very glad to be informed of errors, omissions or misrepresentations.

R. B. Barlow

Department of Pharmacology
Medical School
Bristol.

Foreword

This book gathers together chemical ideas which are important for understanding how drugs act and how new drugs may be developed. Students meet some of these ideas in courses in chemistry, biochemistry and pharmacology but in my experience they find it difficult to put together information often acquired in different years and from different departments; they need it set out in a book.

I believe also that it helps if they can see how these ideas may be applied to numerical problems so I have included examples (with answers), many of which are taken from research work. These should therefore provide a chance for the student both to test his or her own understanding and to see the sort of results which form the bricks of which any experimental science is made. In all the problems the arithmetic takes less than ten minutes with a small calculator which has exponential functions and logarithms.

From this collection of information and ideas the book proceeds to consider some of the attempts which have been made to relate chemical properties quantitatively to biological activity and to predict the activity of new compounds from results obtained with a carefully selected trial group. It is important that the medicinal chemist and the biologist work together in this aspect of drug development and the book attempts to explain what the medicinal chemist is doing and to consider some of the assumptions he makes.

CHAPTER 1

Measuring Drug Activity

'A branch of science comes of age when it becomes quantitative.'
J. H. Gaddum, *Edinburgh Medical Journal* (1942) no. 49, p. 731.

Introduction

The development of drugs starts with the recognition that some
particular pharmacological effect may be useful therapeutically.
Sometimes an application can be seen before there is a compound
with the desired properties. Past history often suggests the reverse:
several important drugs, such as the sulphonamides, have been
made by the chemist years before their therapeutic value was
realised. In either situation an initial discovery depends upon both
an understanding of what may be useful and upon a knowledge of
what drugs do; it requires both the flash of genius and the
humdrum collection of pharmacological information.

Once something potentially useful has been found it is necessary
to know not only what a substance does but how well it does it, so
that comparisons can be made and better drugs discovered. This
may not be easy. The effects which the pharmacologist is called
upon to study are extremely diverse. Some are much more difficult
to detect and compare than others and less reliance can be placed
in the results. The effects of a drug on the growth or function of
cells may be relatively easy to see or to measure and should be
highly reproducible; effects on the central nervous system resulting
in changes in behaviour are likely to be a very different matter.
Drug development depends critically upon adequate testing and
this demands an appreciation of the problems of quantitative
pharmacology.

When a suitable pharmacological test has been devised the
successful development of new drugs depends upon finding what
changes in chemical structure lead to increased activity. It may be
possible from the relation between structure and activity to obtain
some information about how the drugs act, which in turn may
suggest new structures which are worth investigation. This part of

the work is both a scientific problem in chemical ('molecular') pharmacology and of great practical importance, because the making and testing of many new compounds is expensive and takes time. It is an intellectual challenge to know why activity alters with chemical structure but the answers may also make it possible to develop better drugs rapidly and economically.

Because drug development is impossible without proper quantitative pharmacological tests the book begins with a consideration of the problems of measuring biological activity.

Measuring the Effects of Drugs: 'Responses'

If a drug produces an effect, it changes the properties of some of the cells in an organism. Some effects may be actually visible but most can only be detected with suitable apparatus. Many effects involve physical or chemical changes which can be measured. These can be classified into:

  (i) changes in mechanical properties, e.g. in length, volume, pressure, flow, rates of contraction;

 (ii) changes in electrical properties, e.g. in the polarisation or conductance of the cell membrane, potential changes due to the activity of the heart or the central nervous system, changes in rates of firing of neurones;

(iii) chemical changes, e.g. in the concentration of substances involved in cell function, such as nervous transmission, or in metabolism; this may include changes in the amounts of enzymes involved in the synthesis or metabolism of such substances.

Although the apparatus used to detect such effects may be sophisticated, the response is usually the simplest type to consider, because it can be measured and its size depends, within limits, on the dose.

In some tests the effect cannot be measured, perhaps because it is too complex, but it may be possible to devise some scoring system which allows effects to be compared and arranged in order. This often applies in tests for the effects of drugs on behaviour and in the clinical assessment of the treatment of disease.

In other situations, such as toxicity tests, the responses can only be 'all-or-none'. An animal or a cell either lives or dies; a nerve cell either conducts or it does not; a striated muscle cell either twitches or it does not. Responses can be made to appear graded by taking groups of animals or cells or working with a nerve trunk or a muscle bundle, rather than with single cells. Because individuals differ in their sensitivity to a drug there will be a dose range within which only a proportion will be affected, rather than all of them or none at all. The response of a group in this situation is termed quantal. The effect depends on the proportion of units responding in an all-or-none fashion and it varies with the dose because of the differences in the sensitivity of individual units. This is not the same as where the performance of the unit is variable and the response is termed 'graded'. For example effects involving smooth muscle are graded but those involving twitches from striated muscle are quantal. Effects can be classified into:

(i) what can be measured and gives a graded response;
(ii) what can be scored and ranked;
(iii) what can be counted or depends on the proportion of units responding in an all-or-none fashion.

The complexity of the apparatus used to detect the effect is no guide to the complexity of a test. Elaborate equipment is usually employed in order to simplify the response but this may also be simplified by a suitable choice of test preparation. By working with enzymes, single cells, or even pieces of tissue, rather than whole animals it should be possible to simplify the nature of the response by reducing the number of variables on which it depends. In particular it should be possible to reduce, or even eliminate, the complicating effects of time on the response. In a whole animal the effect of a drug usually rises to a peak (depending on how it is administered) and then declines. What happens at any particular moment depends on:

(i) how the drug reaches its site of action;
(ii) what it does when it gets there;
(iii) how it is removed.

These processes may well be quite unrelated to each other. In a test on an isolated preparation with the drug added directly to the bathing fluid it may be possible to see what a drug does without the complications of transport and removal. In some situations it is even possible to make measurements in conditions of equilibrium, or at least in a steady state, and so to measure effects at a much more fundamental level. Such tests, however, are far removed from the clinical situation and there is therefore a conflict between the need to simplify to obtain reliable tests and the difficulty of simplifying tests for properties which are clinically most interesting. In fact a single type of test is not enough; all three processes, transport, action and removal, need separate investigation if the effects in the whole animal or man are to be understood.

Reproductibility of Graded Responses

The biggest problem in all drug testing is the way the response may vary from one experiment to another. It is not enough that the effect of a dose of one compound appears to be bigger than that of a dose of another. It is necessary to ask whether there really is any difference between the two results and to see if the difference is big enough to refute the null hypothesis, that the results could all have come from the same group.

If the test preparation produces a graded response and the variation in response is due only to chance, individual responses should be distributed *normally* about the true value. The distribution is Gaussian and can be seen in Figure 1.1, where the probability, P, of obtaining a result is plotted against its distance from the mean, μ, expressed as a multiple, N, of the standard deviation, σ.

If the same dose has been tested n times and the individual responses are $r_1, r_2, r_3, \ldots, r_n$, the *mean*, $\bar{r} = Sr/n$ where $Sr = r_1 + r_2 + r_3 + \ldots + r_n$. An estimate of the *variance* of the results, $V = Sd^2/(n-1)$, where Sd^2 is the sum of the squares of the deviations from the mean, i.e. $Sd^2 = S(r-\bar{r})^2$. There is actually no need to calculate the deviation of each individual result from the mean because $Sr^2 = S(\bar{r}+d)^2 = n\bar{r}^2 + 2\bar{r}\,Sd + Sd^2$, but $Sd = 0$, so $Sd^2 = Sr^2 - n\bar{r}^2 = Sr^2 - (Sr)^2/n$.

It is therefore only necessary to sum r and r^2, a simple operation for a calculator, and the mean and the variance of the results can be worked out. An estimate of the *standard deviation* of the results, $s = V^{1/2}$. The mean, $\bar{r}$, is an estimate of the true mean, μ; the standard deviation, s, is an estimate of the true standard deviation, σ.

For example if there were six responses: 15, 19, 21, 23, 18, 20, the mean, $\bar{r} = 116/6 = 19.3$; $Sr^2 = 2280$, $Sd^2 = 2280 - 6(19.3)^2 = 2280 - 2242.7 = 37.3$: $V = 7.5$ and $s = 2.7$.

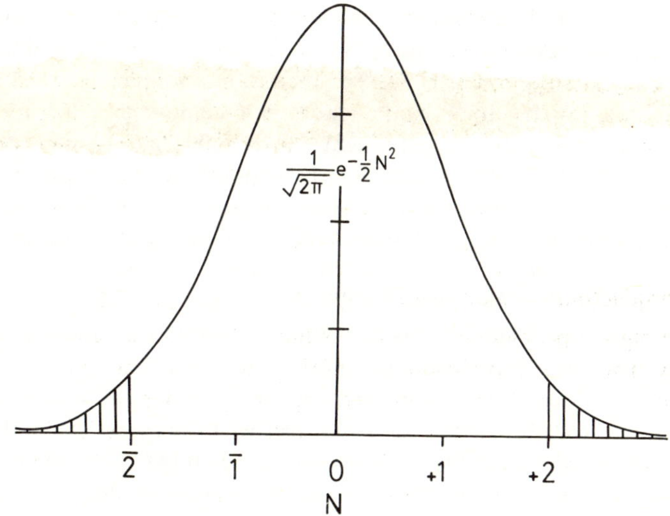

FIGURE 1.1 The graph shows the function

$$\frac{1}{(2\pi)^{1/2}}\,e^{-\frac{1}{2}N^2},$$

i.e. dP/dN, plotted against N. The probability

$$P = \int_{N_1}^{N_2}\left(\frac{1}{(2\pi)^{1/2}}\,e^{-\frac{1}{2}N^2}\right)dN$$

and the integral between $N = -1$ and $N = +1$ encloses 68.2% of the total area, i.e. there are roughly two chances in three that a result will lie within the range $\mu - \sigma$ and $\mu + \sigma$. The integral between $N = -2$ and $N = +2$ encloses 95.5% of the total area so there is only roughly one chance in twenty that a result will lie outside the range $\mu - 2\sigma$ and $\mu + 2\sigma$, which represent the two 'tails' (shaded). For a single tail there is only one chance in forty that a result will lie above $\mu + 1.96\sigma$ and one in forty that it will lie below $\mu - 1.96\sigma$.

If the individual results are distributed normally 68.2% of the values should lie within the range $\mu \pm \sigma$ and 95.5% should lie within the range $\mu \pm 2\sigma$. There is therefore less than one chance in twenty (5%) of obtaining a value outside this range. In the above example two of the results (15, 23) lie outside the range $\bar{r} \pm s$; none lies outside the range $\bar{r} \pm 2s$ (but note that $\bar{r}$ is only an estimate of μ and s is only an estimate of σ).

Our knowledge of the properties of the normal distribution makes it possible to assess the chances of obtaining a particular result, or group of results, and so perhaps to refute the null hypothesis by showing that the chances that the results all come from the same group are too small to be acceptable.

If the example above had included a further result of 28, this would differ from the existing mean by $28 - 19.3 = 8.7$ which is more than $3s$ ($3s = 8.1$). If $\bar{r}$ were really the true mean, μ, and s the true standard deviation, σ, the chances of obtaining such a result would be less than 1%. It might be thought that it really was different from the others and could be rejected but it is necessary to look into the matter more closely. To test the null hypothesis properly the result should have been included with the others in calculating $\bar{r}$ and s; if it is, $\bar{r} = 20.6$ and $s = 4.1$ and the result is only distant from the mean by $(28-20.6)/4.1 = 1.8s$. Moreover, we must make allowance for the fact that $\bar{r}$ is only an estimate of the mean and s is only an estimate of the standard deviation. This can be done by using 'student's t-deviate' (Table 1.1). How good the estimates are will depend on the number of results from which they have been calculated or, more exactly the number of estimates of scatter, called the number of *degrees of freedom*, which is $(n-1)$. (With a mean value and $(n-1)$ results, the last result can be calculated.) On the normal curve 95% of the results lie within the range mean ± 1.96 times the standard deviation (95.5% within $\pm 2\sigma$), but if we are using estimates instead of true values, the number must be replaced by a bigger value, t, whose size depends on the number of degrees of freedom. In the example there are six degrees of freedom and the value of t is 2.45. The expectation is, therefore, that 95% of the results lie within the range $20.6 \pm 2.45 \times 4.1$, i.e. between 10.6 and 30.6, so the extra result (28) could well belong to the same group as the others. It may be different from the others but the evidence is not convincing.

The choice of probability which is considered acceptable depends upon the circumstances. If the probability of obtaining a result is less than one in twenty ($P < 0.05$) it is often regarded as being significantly different but there may be situations where it is desirable to set the limit at one in one hundred ($P < 0.01$). On the normal curve this includes all values in the range $\mu \pm 2.58\ \sigma$; values of t for this level of probability are included in Table 1.1.*

To compare different drugs it is clearly necessary to work with groups of results: single results are of little value. For instance the original six responses, with a mean of 19.3 and a standard deviation of 2.7, might have been obtained with a dose of one drug and in the same conditions a dose of another drug might have produced responses which were: 25, 26, 32, 29, 25, 31 (mean $= 28$; $s = 3.1$).

*The statistical Tables 1.1 to 1.4 will be found at the end of this chapter, on pp. 36–8.

Could these all belong to the same group? The mean values differ by $28 - 19.3 = 8.7$: at what level of probability would this difference be significant?

If individual responses are distributed normally, the means of groups of results will also be distributed normally though they will be gathered much more closely around the true value. The standard deviation of the distribution of means is estimated by the *standard error* of a group of results, $se = s/n^{1/2}$. The values for the two sets of results are $se_1 = 2.7/6^{1/2} = 1.1$ and $se_2 = 3.1/6^{1/2} = 1.3$. If two means have true standard deviations σ_1 and σ_2, the standard deviation of the difference between two means is:

$$\left(\frac{\sigma_1^2}{n_1} + \frac{\sigma_2^2}{n_2}\right)^{1/2}$$

where n_1 and n_2 are the numbers of responses from which the means are calculated. An estimate of this standard deviation is:

$$\left(\frac{s_1^2}{n_1} + \frac{s_2^2}{n_2}\right)^{1/2} = (se_1^2 + se_2^2)^{1/2} = (1.1^2 + 1.3^2)^{1/2} = 1.7.$$

The difference between the means is therefore $8.7/1.7 = 5.1$ times the estimate of the standard deviation. There are five estimates of scatter from each group, making ten degrees of freedom in all. The corresponding value of t with $P = 0.05$ is 2.23 and with $P = 0.01$, t is 3.17. The difference is therefore more than would be expected even at the '1 %' level of probability.

The example given above shows the value of working with groups of responses. If only three responses were obtained in the second group, 25, 25, 26, the difference between the means is $25.3 - 19.3 = 6.0$. The standard error of the second group is 0.3 and the estimate of the standard deviation of the difference is $(1.1^2 + 0.3^2)^{1/2} = 1.15$ and the difference between the means is $6.0/1.15 = 5.2$ times the estimate of the standard deviation. With seven degrees of freedom the value of t with $P = 0.05$ is 2.36 and with $P = 0.01$ t is 3.50 so the difference is significant even at the more rigorous level. Although the difference between the means is less than with all six responses in the second group, the standard error is now very small. Indeed the scatter appears suspiciously different in the two groups, though the experiment is poorly designed for testing this because there are only three responses in the second group compared with six in the first.

To investigate the matter, an analysis of variance can be made.

For two groups, size n_1 and n_2, with means $\bar{r}_1$ and $\bar{r}_2$ and an overall mean $\bar{r}$,

$$S(r-\bar{r})^2 = S(r_1-\bar{r}_1)^2 + S(r_2-\bar{r}_2)^2 + n_1(\bar{r}_1-\bar{r})^2 + n_2(\bar{r}_2-\bar{r})^2.$$

(These can easily be calculated from values of Sr and Sr^2, see above.) The first two terms estimate the variance within the groups; the last two represent the variance between the two means. If they are not very different $\bar{r}_1$, $\bar{r}_2$ and $\bar{r}$ will be close together and these last two terms will be small. For the two groups, 15, 19, 21, 23, 18, 20 and 25, 25, 26, the actual figures are:

$$110.1 = 37.3 + 0.7 + 6(19.3 - 21.3)^2 + 3(25.3 - 21.3)^2$$

$$= 38.1 + 24 + 48 \ (=110.1)$$

This information can conveniently be arranged:

	ss	df	V
Between groups	72.0	1	72.0
Within group 1	37.3	5	7.46
group 2	0.7	2	0.35
Total	110.1	8	13.76

(where *ss* indicates the sums of the squares of the deviations, *df* is the number of degrees of freedom and V is the estimate of the variance. Note that for the variance between two means there is only one degree of freedom).

The variance within the first group is much bigger than the variance within the second and the ratio is $7.48/0.35 = 21.4$. It is possible to calculate the limiting value of this variance ratio, F, consistent with both sets of values having the same distribution. With $P = 0.05$, and the greater estimate based on five degrees of freedom and the lesser estimate based on two degrees of freedom, the limiting value of F is 39.3 (Table 1.2A), so it is not possible to confirm the suspicion. Table 1.2A is based on a two-tailed distribution, where no assumptions are made about which may be the greater variance. If it is known from the experimental plan that one should be bigger than the other, the test involves a one-tailed distribution (Table 1.2B) and the ratio would be significant with $P = 0.05$ (F is 19.3). This is not known, however, but even so it is doubtful whether we can assume that the variances are the same and can be pooled and used in a t-test.

If the group of three results with the second compound is

replaced by the original group of six values, 25, 26, 32, 29, 25, 31, the overall mean $\bar{r} = 23.7$, $\bar{r}_1 = 19.3$, $\bar{r}_2 = 28.0$, and the sums of squares of the deviations (ss), number of degrees of freedom (df), estimates of the variance (V) and the variance ratios (F) are:

	ss	df	V	F
First group	37.3	5	7.5	3.76
Second group	48.0	5	9.6	2.95
Total	310.5	11	28.2	

The values of F indicate that the variance in either of the two groups cannot be distinguished from that of all the results taken together. The total 'within sample' variance is $(37.3 + 48.0)/10 = 8.5$. The 'between sample' variance is $n_1(\bar{r}_1 - \bar{r})^2 + n_2(\bar{r}_2 - \bar{r})^2$ divided by the number of degrees of freedom, which is only 1 because we are considering the means of the two groups, so this is $6 \times 4.3^2 + 6 \times 4.3^2 = 221.9$. We have therefore:

		df	F
Between sample variance	221.9	1	
Within sample variance	8.5	10	$\dfrac{221.9}{8.5} = 26.1$

With one degree of freedom for the greater estimate and ten degrees of freedom for the lesser estimate the limiting value of F, $P = 0.05$ is 6.94 (Table 1.2A) so the analysis of variance establishes that the difference between the means is greater than would be expected on the null hypothesis at this level of probability. If it is known that the results in one group should be bigger than those in the other, for instance if they were obtained with the same compound but with a larger dose, the test involves a one-tailed distribution and the variance ratio is significant even with $P = 0.01$ (Table 1.2B).

An analysis of variance would be expected to confirm the findings of a t-test but it is more rigorous, because it checks the assumptions that the sample variance is indistinguishable in both groups.

EXAMPLE 1:1.

Convince yourself that $Sd^2 = Sr^2 - n\bar{r}^2$ by checking, calculating the values for $r = 3, 4, 5, 6, 7$.

Example 1:2.

The following estimates of log affinity constant (see page 158) for acetylcholine receptors in the guinea-pig ileum were obtained:

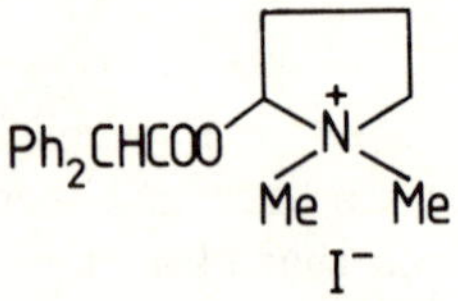

Ph_2CHCOO ... N^+ (Me, Me) I^-	Ph_2CHCOO ... N^+ (Me, Me) I^-
6.974	7.090
6.984	7.083
6.999	7.093
6.986	7.072
6.985	7.117
7.004	7.107
7.003	

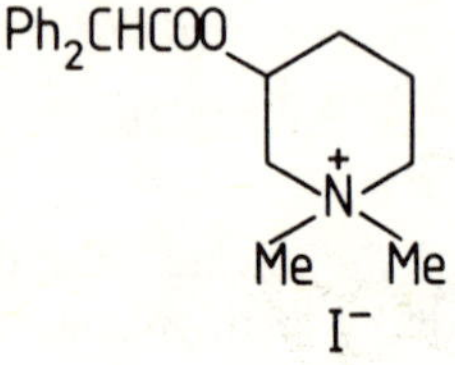

Is there any evidence that the compounds have different affinities?

(From results of Abramson, PhD thesis, Edinburgh University, 1964.)

Ranked Responses

If the responses are not graded but can only be ranked, there is no reason to assume that they will be normally distributed. For instance, the steps in the scale used for scoring may be unequal. If the scores for two drugs were respectively 17, 18, 20, 21, 24 and 20, 23, 25, 26, 28, the values for the middle member of each set, called the *median*, are 20 and 25: 60% of the scores lie between 18 and 21 for the first group and between 23 and 26 for the second. The *interquartile* range, which gives the limits within which half the results lie is sometimes quoted as an indication of the scatter, but these particular results cannot satisfactorily be divided into quarters.

Could these results all belong to the same group? If they are arranged in rank order with the largest $=1$, the values become: 10,

9, $7\frac{1}{2}$, 6, 4: total $(R_1) = 36.5$ for the first and $7\frac{1}{2}$, 5, 3, 2, 1: total $(R_2) = 18.5$ for the second. There are two scores of 20 so the 7th and 8th places tie and both are ranked at $7\frac{1}{2}$. The null-hypothesis may now be tested by the Mann-Whitney (Wilcoxon) procedure. The statistic $U_1 = n_1 n_2 + \frac{1}{2}n_2(n_2+1) - R_2$ and $U_2 = n_1 n_2 + \frac{1}{2}n_1(n_1+1) - R_1$ where n_1 and n_2 are the sample sizes and R_1 and R_2 the corresponding rank sums. So $U_1 = 25 + 15 - 18.5 = 21.5$ and $U_2 = 25 + 15 - 36.5 = 3.5$ (note that $U_1 + U_2 = n_1 n_2$).

Tables of U have been drawn up for values of n_1 and n_2 which indicate the limiting values at various levels of probability. For $n_1 = n_2 = 5$ and $P = 0.05$ the value of U is 2 (Table 1.3) and the criteria for significance are that either U_1 or U_2 must be less than or equal to this value. The results therefore do not indicate any difference at this level of probability.

If an extra result had been obtained with the first drug which was less than 20, the smallest value obtained with the second drug, the total R_1 becomes $36.5 + 11 = 47.5$ and $U_1 = 30 + 21 - 47.5 = 3.5$: the value of U for $n_1 = 6$, $n_2 = 5$ and $P = 0.05$ is 3 so the results are still not significant. If another similar result were obtained, however, $U_2 = 35 + 28 - 59.5 = 3.5$ and the limiting value for $n_1 = 7$, $n_2 = 5$ and $P = 0.05$ is 5, so these results would be significantly different at the level of probability.

This test is described as nonparametric and does not assume the normal distribution. As a result the results contain less information and it is often necessary to make more experiments to reach a conclusion. If the numbers in this example represent scatter which is normally distributed, the two groups of five results can be seen to be significantly different $(P = 0.05)$. The mean and standard errors are 20.0 ± 1.2 and 24.4 ± 1.4; the estimate of the standard deviation of the difference between the means is 1.9 and t for eight degrees of freedom, $P = 0.05$, is 2.3.

A test based on ranking, however, is very useful in a situation where individual results may be zero or infinite, for example in experiments involving measuring sleeping time. Sometimes an animal does not sleep at all, sometimes one dies. These results could not be used in a test based on the normal distribution and suggest that there is really no reason to believe that sleeping times are normally distributed. Such events can still be handled in a test based on ranking, though clearly these extreme results are not as informative as intermediate values.

EXAMPLE 1:3.

In a mock driving test the time taken by an individual to react to a signal was measured. This was repeated 30 minutes after the individual had been given a dose of 0.15 ml ethyl alcohol per kilogram body weight; in another experiment the individuals were given a non-alcoholic 'placebo' instead of the ethyl alcohol. The reaction times (in milliseconds) for 10 males and 10 females were:

Placebo		**0.15 ml ethyl alcohol/kg**	
Before	30 min after	Before	30 min after
Males			
235	245	260	320
275	250	320	380
285	265	280	350
270	295	310	360
290	305	310	270
375	300	285	310
310	270	255	280
265	275	280	310
310	250	250	320
320	330	260	300
Females			
375	300	380	400
285	300	245	370
315	345	350	350
295	245	325	355
320	325	345	305
290	265	295	305
305	335	280	290
260	240	290	320
335	305	270	270
330	330	300	285

Use the Mann-Whitney test to investigate:
1. whether there is a difference between the control responses of males and females;
2. whether the placebo has any effect;
3. whether the alcohol has any effect;
4. whether the effects of alcohol, assessed by the ratio of the reaction times ($t_{\text{alcohol}}/t_{\text{control}}$), are different in males and females.

(From results obtained by Dr P. V. Taberner.)

Graded responses deal with some measurable property and results are adequately described by the mean, number of estimates, and either the variance, standard deviation or standard error. These could be used to calculate the *confidence limits* of the mean, $m \pm t \times se$, within which the true value is expected to lie at a particular level of probability. For example, there are nineteen chances out of twenty that the true means in the worked example above lie within $20.0 \pm 1.2 \times 2.78$ (16.7 and 23.3) and within $24.4 \pm 1.4 \times 2.78$ (20.5 and 28.3) and less than one chance in twenty that they lie outside these limits. This is an impressive way of expressing the results but the number of results must also be stated if any comparison is to be made with other means. Comparisons between means not obtained in the same experiments, however, will only be valid if the effect being measured is something absolute, such as an equilibrium constant, and this is not often true. With results which are scored and ranked such comparisons are totally unjustified. It is usual to quote median values, numbers of results and perhaps the interquartile range to give a general indication of the score and scatter. Actual comparisons require all the individual scores in order to compile the rank order.

Quantal Responses

If the responses are quantal and made up of a number of all-or-none events it becomes necessary to be able to compare the percentage affected in two groups to see whether there is any difference between the responses they represent. This is a problem often encountered, for instance, in trying to decide whether a drug treatment is effective. For example, during the winter term 40 students in a class of 100 took a course of vitamin tablets: 22 of them caught a cold, whereas 45 of the 60 other members of the class caught a cold. Does this indicate that the treatment is beneficial? In these conditions the overall expectation of catching a cold in this group is $(22 + 45)/(40 + 60) = 67\%$, and the expectation of escaping without a cold is 33%. The actual results and the expected results can therefore be set out:

| | | Colds | | No colds | |
Group	Number	observed	expected	observed	expected
With treatment	40	22	26.8	18	13.2
No treatment	60	45	40.2	15	19.8
Total	100	67		33	

If we assume the null hypothesis, that there is no difference between the groups, the discrepancy between observed and expected should be due only to chance. This discrepancy can be estimated by χ^2, which is the sum of

$$\frac{(\text{observed value} - \text{expected value})^2}{\text{expected value}};$$

this can be regarded as a measure of the scatter in relation to the size of the effect. For these results,

$$\chi^2 = \frac{(4.8)^2}{26.8} + \frac{(4.8)^2}{13.2} + \frac{(4.8)^2}{40.2} + \frac{(4.8)^2}{19.8} = 4.342.$$

If the agreement between observed and expected were perfect, χ^2 would be 0 but this is unlikely even with two samples from the same population. The limiting values of χ^2 have been calculated which are consistent with chance variation at various levels of probability and numbers of degrees of freedom. The results in this problem have only one degree of freedom because from the overall expectancy of a cold (67%) and the observed proportion in one of the groups, the proportion in the other group can be calculated and also the proportions of those who do not have a cold. For one degree of freedom the limiting value of χ^2 corresponding to $P = 0.05$ is 3.841, so the results indicate a significant difference at this level of probability, though not at $P = 0.01$ for which $\chi^2 = 6.635$. This test indicates that there is probably a real difference between the incidence of colds in the treated students (55%) and the incidence in untreated students (75%).

When quantal responses are obtained the variation in the proportion affected depends on the variation in the sensitivity of individual units. There is some evidence that the doses producing an all-or-none effect, such as death, are normally distributed, but it is usually assumed that it is the logarithm of the dose which is distributed normally.[1] It is easier, therefore, to consider the variation in the dose producing the effect in a particular proportion of the population (e.g. in 50%) rather than to consider the variation in the effect produced by a particular dose. The assessment of drug activity of all types so far considered, in fact, depends on comparing doses producing comparable responses, though it may be necessary to demonstrate that the test method is sensitive by showing that an increase in dose produces a significant increase in response.

Assessing Biological Activity

If the chemical events producing the responses are understood (these are discussed later in this book) it may be possible to express the effect of a drug in terms of some chemical property, such as an affinity constant, which can be compared with that of another drug. With responses from animals or tissues, however, the mechanism of the response is complex and usually not fully understood. It is necessary to assess activity by comparing the doses of drugs which produce the same biological response and it is not correct to compare the size of the responses produced by the same doses of drugs. Doubling the dose, for instance, does not double the response. The increase varies greatly from one type of test to another and may even vary from one test preparation to another taken from the same animal.

This is well known to pharmacologists who are concerned with bioassay. Many drugs which have useful biological properties cannot be estimated chemically and the potency of a new batch has to be measured by comparing its biological activity with that of a standard sample. If a dose of the new batch produces twice the response obtained with the same dose of the standard preparation, it is not twice as active. There is no reason why twice the dose of the standard should produce twice the response. The relation between dose and response must be determined experimentally. When this has been done, it is possible to calculate what dose of the new batch would produce the same response as the dose of the standard.

Methods of bioassay developed for comparing standard and test preparations containing the same active drug are an obvious starting point for comparing the activities of two different drugs. This extension involves important assumptions, however, which are discussed after an account of bioassay.

Bioassay—Graded Responses

It is usually found that the graph of response against the logarithm of the dose is S-shaped, but that it is approximately linear in the middle range, where the responses are neither very small nor very large. The aim of an assay is to find the dose of the test solution which produces the same response as a particular dose of the standard solution. It is most unlikely that it will be possible to find doses which produce responses which match exactly but it should not be difficult to find doses which produce responses which match approximately. If the dose of standard material produces an effect which is recorded as 30 units (measured, perhaps, by the movement of the pen of a potentiometric recorder) and half this dose of the test solution produces 27 units, the test solution must be nearly twice as active as the standard. To obtain an exact estimate we need to know what the difference of 3 units in the size of the response represents in terms of a change in the dose. If twice the dose of standard has also been tested and produces a response of 70 units, we know that doubling the dose, i.e. increasing log dose by 0.301, increases the response by 40 units. The difference in response of 3 units therefore corresponds to a difference in log dose of $\frac{3}{40} \times 0.301 = 0.023$. To match the effect of the dose of standard exactly the log dose of the test would have to be increased by this amount, i.e. the dose of the test would have to be multiplied by its antilog $= 1.053$. The test solution is therefore estimated to be $\frac{2}{1.053} = 1.90$ times as strong as the standard.

This is an estimate and to assess its reliability we need to know whether the responses to a dose are consistent or not. This necessitates testing the doses several times so the numbers 27, 30, 70 would all be mean values whose variance was known. The procedure is known as a three-point assay and if the doses used are s_1, s_2 and t, and the mean responses they produce are $\bar{S}_1$, $\bar{S}_2$ and $\bar{T}$, respectively, the potency of the test solution can be written

$$\frac{s_1}{t} \times \text{antilog}\left(\frac{\bar{T} - \bar{S}_1}{\bar{S}_2 - \bar{S}_1} \times \log\frac{s_2}{s_1}\right).$$

A three-point assay is occasionally useful when the test preparation is in very short supply. There are great advantages, however, in using two doses of the test solution, t_1 and t_2, as well as of the standard and making a four-point assay. If doubling the dose of standard produces a response which is still in the middle range on the linear part of the log dose response curve, doubling the dose of test should do the same, because both solutions contain the same drug; the ratio t_2/t_1 should be the same as s_2/s_1. There are now two estimates of the slope of the linear part of the log dose response line,

$$\frac{\bar{T}_2 - \bar{T}_1}{\log\left(\frac{t_2}{t_1}\right)} \quad \text{and} \quad \frac{\bar{S}_2 - \bar{S}_1}{\log\left(\frac{s_2}{s_1}\right)}$$

and there are likewise two estimates of the 'effect', i.e. the difference in response $(\bar{T}_1 - \bar{S}_1)$ and $(\bar{T}_2 - \bar{S}_2)$. Suppose that $\bar{S}_1 = 30$, $\bar{S}_2 = 70$,

$\bar{T}_1 = 27$, $\bar{T}_2 = 63$ and that

$$\frac{s_2}{s_1}\left(=\frac{t_2}{t_1}\right) = 2.$$

The average difference in response is

$$\frac{\bar{T}_1 - \bar{S}_1 + \bar{T}_2 - \bar{S}_2}{2} = \frac{27 - 30 + 63 - 70}{2} = -\frac{10}{2} \text{ units.}$$

The average effect of doubling the dose is to increase the response by

$$\frac{\bar{S}_2 - \bar{S}_1 + \bar{T}_2 - \bar{T}_1}{2} = \frac{70 - 30 + 63 - 27}{2} = \frac{76}{2}$$

so the difference in the responses tells us that to obtain an exact match, we should have to increase the log dose of test by

$$\frac{10}{2} \times \frac{2}{76} \times \log 2 = 0.132 \times 0.301 = 0.040,$$

i.e. the dose of test should be multiplied by 1.096: if the doses of test used were half those of the standard, the activity of the test is $\frac{2}{1.096}$ times that of the standard. If we think in terms of the dose of standard which would match the effect of test, log dose should be altered by -0.040 so the dose of standard should be multiplied by 0.912, which comes to the same thing ($0.912 = \frac{1}{1.096}$). If, as in the three-point assay, the doses of test were half those of standard, its activity $= 2 \times 0.912 = 1.824$ times that of the standard. In general terms the potency of the test can be written:

$$\frac{s_1}{t_1} \times \text{antilog}\left[\left(\frac{\bar{T}_2 - \bar{S}_2 + \bar{T}_1 - \bar{S}_1}{\bar{T}_2 - \bar{T}_1 + \bar{S}_2 - \bar{S}_1}\right) \times \log\left(\frac{s_2}{s_1}\right)\right].$$

Note that the number to be 'antilogged' should be close to zero and will be positive if the effects produced by the test solution are greater than those produced by the standard and negative if the effects of the test solution are smaller than those of the standard. A large number to be 'antilogged' indicates a very bad experimental match (or an error in the calculations). Note also that sometimes the effects of low doses may appear to contradict those of high doses. It is possible to have results such as $\bar{S}_1 = 30$, $\bar{S}_2 = 63$, $\bar{T}_1 = 27$, $\bar{T}_2 = 70$. This usually indicates a close match between the responses to test and standard; the net difference in response is

$$\frac{30-47+63-70}{2} = \frac{3-7}{2} = -\frac{4}{2}.$$

The estimates of the slope will be appreciably different, indicating the variability of the responses.

In fact it is not enough merely to be able to estimate the potency of a test solution, it is necessary to assess how reliable the estimate is likely to be. In general the best results will be obtained when the responses to test and standard are closely matched. The bigger the calculated correction factor, the more inaccurate the assay is likely to be, because it assumes that the size of the response is directly proportional to log dose and this is only true over the middle range of responses. The slope should be the same for test and standard solutions and although the same drug is the active agent in each, a bad match may result in a group of responses being outside the range and the lines not being parallel. To check for parallelism and to set limits to the estimate of potency it is necessary to examine the scatter of the results about the mean values from which the estimate of potency has been calculated. The following method has been described for routine bioassay.[2]

In a properly balanced experiment, each of the means, $\bar{S}_1$, $\bar{S}_2$, $\bar{T}_1$, $\bar{T}_2$, is derived from the same number of estimates, n, and the estimate of the variance is:

$$V = \frac{S(S_1 - \bar{S}_1)^2 + S(S_2 - \bar{S}_2)^2 + S(T_1 - \bar{T}_1)^2 + S(T_2 - \bar{T}_2)^2}{4n(n-1)}.$$

If n is not the same for all groups but the numbers n_1, n_2, n_3, n_4, differ only slightly the denominator may be replaced by $(n_1 + n_2 + n_3 + n_4 - 4)(n_1 + n_2 + n_3 + n_4)/4$.

The difference in effect due to the size of the dose $E = (\bar{S}_2 - \bar{S}_1 + T_2 - T_1)/2$ and its variance $V(E)$ is the same as V above. The slope of the log dose response lines, $b = E/I$ where $I = \log(s_2/s_1)$ and the variance of the slope, $B = V/I^2$. The index of significance of the slope, $g = Bt^2/b^2$ where t is the value of student's parameter for the chosen level of probability and the number of degrees of freedom, $4(n-1)$ (or $n_1 + n_2 + n_3 + n_4 - 4$).

The effect difference, F, due to the difference between standard and test solutions is $(\bar{T}_1 - \bar{S}_1 + \bar{T}_2 - \bar{S}_2)/2$ and M, the log of the potency ratio is F/b. The variance of the effect difference $V(F)$, is the same as the pooled estimate, V, and the confidence limits of M are:

$$M + \frac{gM}{1-g} \pm \frac{t}{b(1-g)}(A(1-g)+BM^2)^{1/2}$$

$$= \frac{M}{1-g} \pm \frac{t}{b(1-g)}(A(1-g)+BM^2)^{1/2}$$

where $A = V(F) = V$; if g is small, e.g. less than 0.1, the expression approximates to

$$M \pm \frac{t}{b}(A+BM^2)^{1/2}.$$

This gives the range within which the true value of M is expected to lie at a particular level of probability. The limits to the estimate of potency can be calculated by taking antilogs. Note that if these are thought of simply as the correction factor for M they are:

$$\frac{gM}{1-g} \pm \frac{t}{b(1-g)}(A(1-g)+BM^2)^{1/2}.$$

To test whether the slopes of standard and test are different it is necessary to calculate the variance of the effect difference, G, due to differences in the slopes: $G = (\bar{T}_2 - \bar{T}_1) - (\bar{S}_2 - \bar{S}_1)$ and $V(G)$ is $4V$. For the slopes to be indistinguishable, $G/[V(G)]^{1/2} = G/2V^{1/2}$ should not exceed the value of student's t for the level of probability selected and the appropriate number of degrees of freedom, $4(n-1)$ or $(n_1 + n_2 + n_3 + n_4 - 4)$.

If only a three-point assay is performed with two doses of standard but only one of test, the pooled estimate of the variance of the means, V, is:

$$\frac{S(S_1 - \bar{S}_1)^2 + S(S_2 - \bar{S}_2)^2 + S(T - \bar{T})^2}{3n(n-1)}$$

(the denominator is $(n_1 + n_2 + n_3 - 3)(n_1 + n_2 + n_3)/3$ if incomplete groups have been used). The variance of the effect difference, $V(E)$, is $2V$ and the variance of the slope, B is $2V/I^2$ (these are based on responses only to one pair of doses). The variance due to the use of the test solution, $V(F)$ is $3V/2 = A$ and the fiducial limits can be calculated as in a four-point assay.

Suppose, for example, that the mean values shown above were calculated from the following individual responses:

$$
\begin{array}{llll}
 & & & Sd^2 \\
S_1 & 31, 28, 34, 27 & \bar{S}_1 = 30 & 30 \\
S_2 & 64, 75, 67, 74 & \bar{S}_2 = 70 & 86 \\
T_1 & 22, 23, 29, 34 & \bar{T}_1 = 27 & 94 \\
T_2 & 65, 67, 59, 61 & \bar{T}_2 = 63 & 40
\end{array}
$$

Values of Sd^2, the sums of the squares of the deviations, are calculated as usual from $Sr^2 - n\bar{r}^2$ and

$$
V = \frac{30 + 86 + 94 + 40}{4 \times 4 \times 3} = \frac{250}{48} = 5.21.
$$

The effect difference due to the difference in the slopes,

$$
G = (63 - 27) - (70 - 30) = (-)4
$$

and

$$
\frac{G}{2V^{1/2}} = \frac{4}{2(5.21)^{1/2}} = 0.876
$$

which is much less than t for twelve degrees of freedom (which is 2.18, $P = 0.05$), so there is no significant difference in the slopes of the two lines. The slope, b, and the log of the potency ratio, M, have already been calculated

$$
b = \frac{76}{2\log 2} = 126.2; \quad M = \frac{-10}{2} \times \frac{2\log 2}{76} = -0.040
$$

and the variance of the slope,

$$
B = \frac{5.21}{(\log 2)^2} = 57.5.
$$

The index of significance,

$$
g = \frac{57.5(2.18)^2}{(126.2)^2} = 0.017,
$$

which is negligible, so the log limits are:

$$
M \pm \frac{2.18}{126.2}[5.21 + 57.5(-0.040)^2]^{1/2}
$$

$$
= \pm 0.0173(5.21 + 0.092)^{1/2}
$$

$$
= \pm 0.0398
$$

and the limits are 0.912 and 1.096 times the estimate, i.e. between 0.912×1.824 $(=1.66)$ and 1.096×1.824 $(=2.00)$.

If only a three-point assay had been made, with the results from t_2 omitted, the values become

$$V = \frac{30+86+94}{3\times4\times3} = \frac{210}{36} = 5.83; \quad b = \frac{70-30}{\log 2} = 132.9;$$

$$M = -0.023; \quad B = \frac{2\times5.83}{(\log 2)^2} = 128.7;$$

$$g = \frac{128.7(2.26)^2}{(132.9)^2} = 0.037,$$

which is again negligible so the log limits are:

$$M \pm \frac{2.26}{132.9}[8.75+128.7(-0.023)^2]^{1/2}$$

$$= \pm0.0172(8.75+0.068)^{1/2} = \pm0.0505$$

and the limits are 0.890 and 1.123 times the estimate.

In many four-point assays the limits should be better than $\pm10\%$, which is the usual requirement for pharmacopoeial standards. It is much easier to achieve with some types of preparation which give much more consistent responses than others. Results of this quality, for instance, can often be obtained by students testing unknown solutions of acetylcholine-like compounds on a piece of guinea-pig intestine. Not only are the limits usually within the range $\pm10\%$ but so is the agreement with the concentration as noted by the person who prepared the test solution. This tissue is particularly easy to work with but, even so, allowance must be made for changes in sensitivity during the assay. It is common to observe a definite trend, with the tissue becoming more (or less) sensitive, so it is important to arrange that the doses, s_1, s_2, t_1, t_2, are evenly distributed throughout the assay. The order in which they are given should also be varied because the response to one dose may be affected by the previous one. It is usual therefore to give the doses in an order based upon a Latin square, in which each of the 4 doses, A, B, C, D appears once in each line and once in each column, e.g.

$$
\begin{array}{cccc}
A & B & C & D \\
B & A & D & C \\
C & D & B & A \\
D & C & A & B
\end{array}
$$

The method described above provides a set of rules which can easily be applied and can be made into a computer programme. The rules are based on an analysis of variance of the responses[3] and, provided the test for parallelism is included to check that it is permissible to set limits, the two methods are really equivalent.

EXAMPLE 1:4.

An extract made from ox adrenal glands produced a rise in the blood-pressure of anaesthetised rats. It was diluted 1 to 50 and the mean rises produced by 0.1 ml and 0.2 ml of the diluted solution were 35 and 64 mm Hg respectively. The mean rises produced by 0.15 ml and 0.3 ml of (-)-adrenaline, 0.3 mM, were 32 and 67 mm Hg, respectively. If the extract contains only (-)-adrenaline, calculate the concentration in the extract.

EXAMPLE 1:5.

A solution of carbachol (A) produced contractions of the guinea-pig ileum which were recorded with a lever writing on a smoked drum. The mean size of the records of the responses to 0.2 ml and 0.4 ml of solution A added to an organ bath with a volume of 10 ml, were 16 and 38 respectively. Mean responses of 18 mm and 41 mm were obtained with 0.15 and 0.3 ml respectively of a 10^{-5} M solution of carbachol added to the bath. What is the concentration of carbachol in solution A?

EXAMPLE 1:6.

In a similar experiment a solution of *n*-pentyl-trimethylammonium iodide was assayed against a standard solution, 5×10^{-5} M, on the guinea-pig ileum and the following responses were obtained:

0.1 ml standard: 17, 16, 19, 25.5
0.2 ml standard: 30.5, 31.5, 38, 42.5
0.15 ml test: 15, 15, 19, 25
0.30 ml test: 29.5, 32, 33, 43

Estimate the potency of the test solution and the 95% confidence limits. The estimate is within 10% of concentration noted by the person who prepared the test solution (from the same stock solution as the standard). Comment on the size of the limits.

EXAMPLE 1:7.

An extract was made from nettle leaves by homogenising 1 g in 100 ml Tyrode's solution. This was diluted 1 to 100 and tested on the isolated guinea-pig ileum. Mean responses to 0.2 and 0.4 ml of this solution were 18 and 44 units, respectively, measured as movement of the pen across the chart-paper of the recorder. Mean responses to 0.15 and 0.3 ml acetylcholine, 5×10^{-6} M, were 21 and 49 units respectively. If the responses were produced by acetylcholine in the extract, calculate the concentration and estimate the minimum concentration in the leaf.

How do these results compare with estimates of an average content of about 0.1 μg acetylcholine chloride for a single hair from nettle stalks and a loss in weight of 4.5 mg when 500 hairs are dried carefully?

In estimates of the acetylcholine content of single hairs the following average values ($\pm$ standard error and number of hairs tested) were obtained: on rabbit intestine, 0.065 ± 0.016 (10) μg with hairs from nettle stalks: on guinea-pig intestine, 0.091 ± 0.035 (10) μg with hairs from nettle stalks: on guinea-pig intestine, 0.039 ± 0.002 (6) μg with hairs from the upper surface of nettle leaves: on guinea-pig intestine, 0.033 ± 0.007 (4) μg with hairs from the lower surface of nettle leaves. Do these figures indicate any significant differences between results obtained with guinea-pig as opposed to rabbit intestine, or from different parts of the plant?

(From results of Emmelin and Feldberg (1947) *J. Physiol.* no. 106, p. 440; see also Edinburgh Staff 'Pharmacological Experiments on Isolated Preparations' (Churchill Livingstone, 1974), experiment 13b, p. 67.)

Bioassay—Quantal Responses

In theory quantal responses could be used to compare the activities of two drugs in exactly the same way as graded responses, by working with groups and observing the proportion in a group affected by the low and high doses of standard and test. The figures in the previous example on page 16 for instance, might refer to the percentage of animals killed in each group, 30% and 70% by the doses of standard and 27% and 63% by the doses of test. Such a method, however, would be extremely wasteful of information because statistical theory can be used to establish the shape of the relationship between log dose and percentage mortality.

Individual animals or cells vary in their sensitivity to a drug. Some are affected by much lower doses than others and the extent of the sensitivity also can vary greatly from drug to drug and with the population on which they are being tested. The range in sensitivity, for instance, might well be different in different drugs if they are producing the effect by different mechanisms. There is some evidence that the variation of the dose producing a quantal response, such as death, is normal (see below) but it is more usual to assume that the log dose is distributed normally. Accordingly, if $\bar{x}$ is the mean effective log dose for the population and σ is the standard deviation of the distribution, the effective log doses for approximately two out of three individuals will lie in the range $\bar{x} \pm \sigma$ and for nineteen out of twenty they will lie within the range $\bar{x} \pm 1.96\sigma$.

When a dose is given to a group of individual animals or cells, it will affect all those sensitive to doses up to and including the value given, so $\bar{x}$ will affect 50%, $\bar{x} - \sigma$ will affect about one out of six (actually 16%), $\bar{x} + \sigma$ will affect about five out of six (84%), $\bar{x} - 2\sigma$ will affect 2.2%, $\bar{x} + 2\sigma$ will affect 97.8% and so on (Figure 1.2). A table can be constructed showing the relation between percentage affected and log dose, expressed as $\bar{x} + n\sigma$, where n is called the *normal equivalent deviation*. Usually 5 is added to n, to make all likely values positive and these numbers are called *probits* (Table 1.4).

If the distribution is normal, the graph of the probit corresponding to the percentage of the group affected plotted against log dose should be a straight line: $\bar{x}$ will correspond to a probit of 5 and values of x separated by 1 probit are an estimate of the standard deviation (the slope of the line is $1/s$). If the dose is used, rather than log dose, $\bar{x}$ is the dose effective in 50% of the animals and is written ED_{50}, or LD_{50} if it is the lethal effect which is being tested. If log doses are used, $\bar{x}$ is $(\log ED)_{50}$, which is not necessarily the same as $\log(ED_{50})$, because the mean will be geometric rather than arithmetic, but the difference is not usually

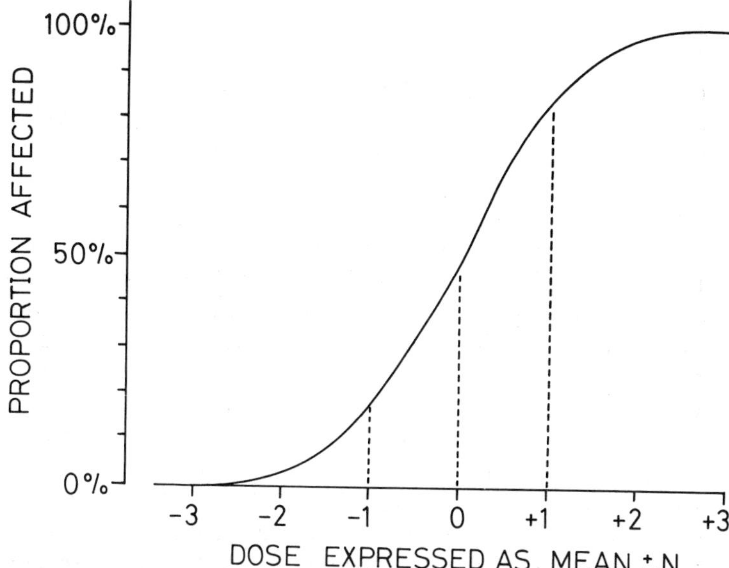

FIGURE 1.2 Integrated form of the normal frequency distribution (Figure 1.1) showing the proportion likely to be affected by a dose above or below the mean by a particular multiple N, of the standard deviation. If the actual dose were plotted the curve would have the same form but the x-axis would be expanded or contracted depending on the scale and the size of the standard deviation, σ. Because the relation between proportion affected and dose is the same however the dose is expressed, the graph of dose against N corresponding to the proportion affected should be a straight line whose equation is dose $= \bar{x} + N\sigma$. Usually 5 is added to N, giving the *probit* of the proportion affected, so that values are all positive.

Note that at either end of the curve large changes in N produce only small changes in the proportion affected. In plotting the graph of dose against probit proportion affected, these results should carry less weight and a table of weights for probits gives values of:

$$P = 3.0 \quad 3.5 \quad 4.0 \quad 4.2 \quad 4.4 \quad 4.6 \quad 4.8 \quad 5.0$$
$$7.0 \quad 6.5 \quad 6.0 \quad 5.8 \quad 5.6 \quad 5.4 \quad 5.2$$
$$W = 0.13 \quad 0.27 \quad 0.44 \quad 0.50 \quad 0.56 \quad 0.60 \quad 0.63 \quad 0.64$$

large. Consider, for example, the following results obtained in a test of the toxicity of the alkaloid (-)-cytisine given orally to mice.

Dose µmol/kg	log dose	Mice killed number	%	Probit
208	2.318	0/10	0	
368	2.566	4/10	40	4.75
544	2.736	5/10	50	5.00
648	2.812	6/10	60	5.25
820	2.914	7/10	70	5.52
1005	3.002	8/10	80	5.84
1309	3.117	10/10	100	

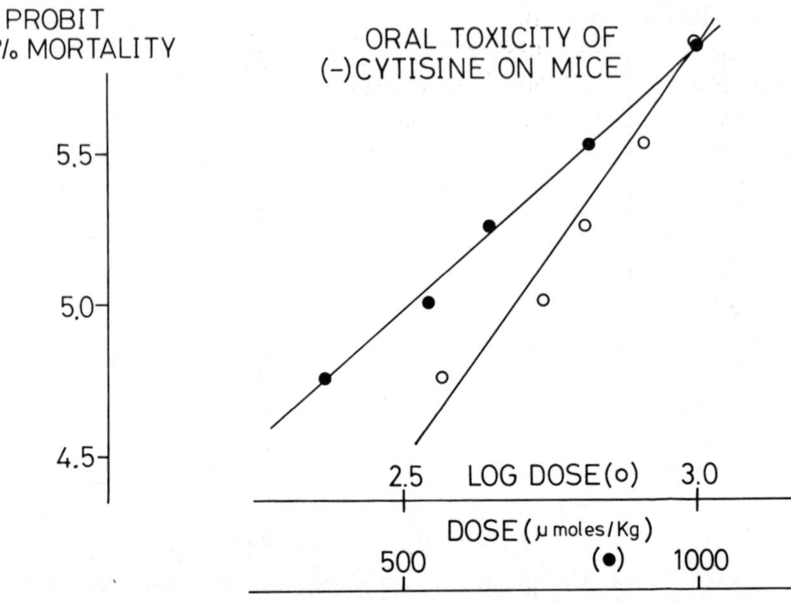

FIGURE 1.3 Toxicity of (−)-cytisine in mice. The probit of the percentage killed is plotted against the dose in μmol/kg (●) and against the log dose (○). The lines have been calculated by the method of least-squares (see Appendix).

(Results of L. J. McLeod, MSc Thesis, University of Edinburgh, 1968.)

Graphs of probit percentage killed against log dose and against dose are shown in Figure 1.3. Either graph could be called a straight line, especially when it is remembered that the response can only alter in quanta of 10%. The lines shown were fitted by the method of least squares (see Appendix) and give values of $\bar{x} = 2.697$, for the graph with log dose, which corresponds to $LD_{50} = 498$, whereas when the dose is used, $LD_{50} = 520\,\mu mol/kg$. The corresponding estimates of the standard deviation are 0.399 and 576. This indicates a reason for working with log dose rather than with dose. There is clearly a big range in the sensitivity of individual mice to this particular compound but the value of s when the dose is used is actually bigger than $\bar{x}$, implying that some of the animals should be sensitive to negative amounts! The distribution, in fact, appears to be skew, rather than symmetrical.

In drawing the straight line fitting the probits to dose or log dose, allowance should be made for the greater variation to be expected with very low or very high proportions killed (see Figure

1.2). Tables of weights for probits have been calculated[2] which make it possible to allow for the greater importance of results in the middle range. With these particular results the weighted fit is not much different ($\bar{x} = 498.5$, $s = 0.403$ with log dose; $\bar{x} = 521$, $s = 576$ with dose). Full details for making estimates of LD_{50} or ED_{50}, including using results of 0% and 100% and the estimation of confidence limits are given by Gaddum[2] and Litchfield and Wilcoxon.[4]

If it is assumed that the probit of the percentage affected is directly proportional to log dose, bioassay calculations used with graded responses can be applied directly to tests involving quantal responses. There is, in fact, no need to limit results to those with responses in the middle range because the graph of probit against log dose should be a straight line over the whole range, though the variation will be greater at the ends.

For example, a dose of a standard tincture of strophanthus killed two out of a group of twenty frogs and 1.5 times this dose killed sixteen out of twenty; a dose of a test extract killed seven out of twenty and 1.5 times this dose killed nineteen out of twenty. The probits corresponding to 10% and 80% are 3.72 and 5.84, respectively, and to 35% and 95% they are 4.61 and 6.64, respectively. The log of the potency ratio is therefore

$$\frac{4.61 + 6.64 - 3.72 - 5.84}{6.64 - 4.61 + 5.84 - 3.72} \times \log(1.5) = \frac{1.69}{4.15} \times 0.176 = 0.0717$$

and the test is estimated to be 1.18 times as active as the standard.

Although the probits corresponding to the responses S_1, S_2, T_1, T_2 are single values, they have been derived from experiments involving a group of animals and the variance of each probit $= 1/\overline{wn}$ where n is the number of animals in the group, w the weight factor for the probit (see above) and $\overline{wn}$ is the average, $Swn/4$. The weight factors for the particular probits are, 0.34 (for 3.72), 0.60 (for 4.61), 0.49 (for 5.84) and 0.22 (for 6.64); note that the weights are greater for probits nearer to 5. All the groups are the same size ($n = 20$) and the average weight is 0.412 so the variance of each probit $V = 1/8.25 = 0.1212$. The effect difference due to the difference in slopes, $G = (6.64 - 4.61) - (5.84 - 3.72) = (-)0.09$ and applying the test for parallelism (page 19),

$$\frac{G}{2(V)^{1/2}} = \frac{0.09}{2(0.348)} = 0.129.$$

With four groups of twenty animals there are 4×19 degrees of freedom and the value of t will be the limit, 1.96 for $P = 0.05$, so clearly the results do not indicate any deviation from parallelism ($1.96 \gg 0.129$), even though they include probits corresponding to results of 10% and 95%. The calculation of limits for the potency estimate then proceeds as on page 19.

$$F = 0.845, \quad E = 2.075, \quad b = \frac{2.075}{0.176} = 11.8,$$

$$B = \frac{0.1212}{(0.176)^2} = 3.91, \quad g = \frac{3.91(1.96)^2}{11.8^2} = 0.108$$

and $M = 0.0717$ so the log confidence limits are:

$$M + \frac{gM}{1-g} \pm \frac{t}{b(1-g)}(A(1-g)+BM^2)^{1/2}$$

$$= \frac{0.0717}{(1-0.108)}$$

$$\pm \frac{1.96}{11.8(1-0.108)} \times (0.1212(1-0.108)-3.91(0.0717)^2)^{1/2}$$

$$= 0.1471 \text{ and } 0.0137$$

and the limits to the estimate of potency ($P = 0.05$) are 1.40 and 1.03. Note the skewness of the limits about the estimate (1.18), associated with the large value of g.

Bioassay—Ranked Responses

For responses which can only be scored and ranked there is no justification for assuming any particular shape for the relation between score and response and such results cannot be treated as if they were graded responses. Some comparison may be possible, however, if care is taken to adjust the doses so that the scores to standard and test match very closely, and only a very small correction need be made, based on the unjustified assumption that score is proportional to log dose.

Example 1:8.

The lethal doses of standard and test solutions of digitalis were determined by infusing them at a standard rate intravenously into anaesthetised cats until the heart stopped; the amount of the particular solution which killed the animal was expressed in ml/kg and the following results were obtained:

Standard solution (1 unit/ml)	18.2, 19.6, 17.0, 19.7, 21.1
Test solution	12.0, 10.0, 8.7, 10.2, 9.6

Calculate the mean lethal doses and the mean log lethal doses and compare the two estimates of the potency of the test solution. Calculate the 95% confidence limits for the estimate of potency based on the difference between mean log lethal doses. (Gaddum, *Pharmacology*, 4th edition, 1953, p. 492, Oxford University Press.)

EXAMPLE 1:9.

In the assay of a test preparation of insulin, mice were taken in groups of 24; convulsions were observed in 17 of the group given 28 milliunits of insulin, in 4 of the group given 16.8 milliunits of insulin, in 19 of the group given the test solution and in 8 of the group given 0.6 of this dose (though all the doses were given in the same volume). Calculate the potency of the test solution and the 95% confidence limits.
(Burn, Finney and Goodwin, *Biological Standardization*, 2nd edition, 1952, p. 211, Oxford University Press.)

Application of Bioassay Methods to the Comparison of Drugs

Although some tests used in bioassay give results with a precision as high as $\pm 10\%$, it does not follow that this will be achieved when the methods are used to compare the activities of different drugs. In bioassay the same drug is present in both the standard and the test solution and by adjusting the doses it must be possible to produce responses which match. When the standard and test solutions contain different drugs there is no reason why this should be so. Time is an important component in many responses and the different drugs may produce effects which have different time-courses. It will then be impossible to produce truly comparable responses, however the doses of the drugs are altered. If the test involves observing peak effects or responses after some particular time, the differences in time-course will be overlooked but, though it will be possible in these circumstances to find doses of the drugs which appear to produce comparable responses, the relative potency may be quite different from what it is in other tests where the responses are observed after some different time interval. There is also no reason why the slopes of the log dose response curves for the two drugs should be similar; if the lines are not parallel, the relative activity of the drugs will depend on the size of the response at which the comparisons are made.

These possible differences between drugs have long been used by pharmacologists in trying to establish the identity of biologically active compounds. If test and standard solutions contain the same substance, the potency ratio should be the same whatever test is used to estimate activity. Bioassays are therefore made on several different types of test system and the estimates of potency should be the same, within the limits of experimental error, if the test drug is identical with the standard with which it is being compared. This technique is known as the method of *parallel assay* and the ratio of the estimates of potency in two different assays is called the *index of discrimination*, which should be 1 if the compounds are not different. It is impossible to establish identity completely in this way, because there is always the possibility that a test on some new preparation may give a result which is significantly different from the others, but after agreement in a number of assays, the chances of such an event may be small, particularly if there has been a wide choice in the type of test preparation used.

When comparing the activities of drugs, therefore, it is necessary to establish:

(i) that the drugs produce responses which have similar time-courses;
(ii) that the drugs have log dose response lines which are reasonably parallel.

It is only when these conditions are fulfilled that it will be possible to express the relative activities of the compounds as a number. If the molar concentrations producing comparable responses are [A] and [B], the *molar activity* of drug B relative to drug A will be [A]/[B] and if B is the more active, it will have the smaller concentration and the ratio will be greater than 1. Alternatively the ratio can be expressed as [B]/[A], the ratio of the concentrations producing comparable effects (the *equipotent molar ratio*), and if B is the more active this will be less than 1. It may be difficult to associate higher activity with a fraction, but this method of expressing the result should appeal to the person who actually does the experiment and weighs out the material and makes up the solutions.

Distinction between Agonists and Antagonists

When comparing effects of drugs an understanding of what the drug does helps greatly in designing a suitable test. A drug may produce its pharmacological effect in two ways, by stimulating cells or by blocking the actions of compounds which themselves do this. With some cells this blocking action may be an active process but it is also possible that it is a passive one. For example adrenaline will cause an increase in the rate of beating and force of contraction of the heart. This is an active process and the drug may be described as an *agonist*. Acetylcholine will slow the rate of beating and decrease the force of contraction so acetylcholine may be described as an antagonist but in fact its effects are brought about by an active process—slowing the heart and decreasing the force of contraction—so it is an agonist although it behaves as a *physiological antagonist* of adrenaline. The effects of adrenaline are also opposed by a substance such as propranolol which is an antagonist of adrenaline and acts passively; the actions of acetylcholine are opposed passively by atropine. For the passive antagonists to be seen to produce an effect, however, the agonists must be present. If atropine is given to heart or gut which has its vagal innervation intact the heart rate will increase, because the effect of the vagal tone is blocked. Similarly if the sympathetic supply is intact the heart rate will decrease in response to propranolol though the size of the effects will depend on the state of the sympathetic tone. If the nerves to the heart have been cut, however, neither atropine nor propranolol will appear to do anything.

The testing of substances which are passive antagonists, therefore can only proceed when the agonist is also given. The effects could be estimated quantitatively as a percentage reduction in the response to a standard dose of agonist, but this is likely to vary from preparation to preparation, depending on variations in the slope of the log dose response curve. A much more satisfactory way for antagonists whose effects are reversible is to increase the concentration of agonist and estimate the *dose-ratio*. If a concentration A of agonist is needed to produce a particular response in the presence of a concentration B of antagonist whereas this same response is produced by a of agonist by itself, the dose-ratio $DR = A/a$.

Many antagonists are not easily washed out, even when

experiments are performed *in vitro*, so it is necessary to obtain control responses to agonist (in the absence of the antagonist) first and then to obtain responses with higher concentrations (A) of agonist in the presence of the antagonist. When these are steady the control responses to agonist alone are used to calculate what concentration of agonist (a) would have given responses of the same size. The experiments are often performed with two concentrations of agonist alone and two concentrations in the presence of the antagonist, so the calculation of the dose-ratio is very similar to that of an activity ratio based on a four-point assay. The values in the example on page 16 might have been obtained in such an experiment; with the responses to 1 and $2\,\mu$M agonist alone being 30 and 70 units respectively and the responses to in- creased agonist $1X$ and $2X\,\mu$M in the presence of the antagonist being 27 and 63 units respectively. The approximate dose ratio is X but this must be corrected for the difference in the responses. Matching responses would be obtained by dividing the concent- ration of the controls by 1.096 (page 17). This means that the antagonist is producing more antagonism than a dose-ratio X and the correct value is estimated to be $1.096X$. Confidence limits could be set to this estimate but it is more common to obtain a number of estimates of dose-ratio in separate experiments and to calculate the standard error and fiducial limits for the mean value of these.

The dose-ratio is a measure of antagonism. It should increase with higher concentrations of antagonist and can in certain circumstances (Chapter 4) be used to calculate an equilibrium constant which is a fundamental measure of activity. This is not usually possible with agonists, however, whose activity can only be expressed by comparing concentrations producing matching responses, i.e. as an equipotent molar ratio or an activity ratio. With tests made on enzymes, rather than cells, however, it is possible to measure parameters, such as Michaelis constants and maximum rates, which have chemical significance. By working with simplified biological systems the information about drug activity can be more precise but may tell you less about what the drug will do in the clinic.

This section on assessing biological activity is intended to indicate the problems and to illustrate what can be done to obtain information. It should be regarded as a supplement to the many excellent books on statistics and bioassay (some are listed on p. 36), not as a substitute for them.

EXAMPLE 1:10.

An extract, thought to contain acetylcholine, was compared with standard solutions of acetylcholine on the guinea-pig ileum and frog rectus preparations. Because the extract was in short supply, it was tested with only one size of dose on the frog rectus (i.e. in a three-point assay) but on the ileum two sizes of dose were used. The following mean responses were obtained, recorded as movements in millimetres of the pen of the potentiometric recorder connected to the lever system.

Frog rectus:

    Acetylcholine, 5×10^{-5} M; 0.1 ml, 31 mm; 0.2 ml, 53 mm.

    Extract, 0.2 ml, 42 mm.

Guinea-pig ileum:

    Acetylcholine, 5×10^{-6} M; 0.1 ml, 25 mm, 0.2 ml, 59 mm.

    Extract diluted 1 to 20; 0.1 ml, 30 mm; 0.2 ml, 54 mm.

Assuming that it is acetylcholine which is present, calculate the concentration in the extract. Do the results support this idea?

EXAMPLE 1:11.

The frog rectus abdominis preparation gives a contracture when treated with acetylcholine and in an experiment the mean responses to 2×10^{-6} and 4×10^{-6} M were 22 mm and 57 mm respectively. The preparation was then tested in the presence of $(+)$-tubocurarine chloride, 10^{-6} M, and the mean responses to 8×10^{-6} and 1.6×10^{-5} M acetylcholine were 19 mm and 51 mm respectively. What is the dose-ratio produced by this concentration of $(+)$-tubocurarine chloride?

EXAMPLE 1:12.

Carbachol (like acetylcholine) slows the rate of beating of isolated guinea-pig atria and the effect can be expressed as the percentage increase in the time required for 50 beats. In an experiment the mean increases produced by 2×10^{-7} and 4×10^{-7} M were 7.3 and 33.4% respectively. In the presence of diphenylacetylpseudotropine methiodide, 5×10^{-7} M, the mean responses to 4×10^{-6} and 8×10^{-6} M carbachol were increases of 8.0 and 22.8% in the time required for 50 beats. Calculate the dose-ratio.

As has already been mentioned (page 3), what a drug does depends on how it reaches its site of action, what it does when it gets there, and how it is removed. To some extent the processes of transport and removal can be recognised as problems of diffusion and partitioning, because the body is not a single phase but can be regarded as a number of compartments linked by circulating liquid. The most fundamental effect of a drug, what it does at its site of action, is likely to be more complex. Some idea of the types of process which may be involved can be obtained by considering what is known about the chemistry of active drugs.

Types of Drug Action

There are two ways in which drugs appear to be able to affect cells. One type is represented by the compounds described as general anaesthetics which include a large number of substances with very different chemical structures, ranging from organic compounds such as chloroform and ether to inorganic compounds such as nitrous oxide and inert gases such as xenon. Many are gases or volatile liquids which can be inhaled and so act quickly and, perhaps not surprisingly, the rate at which they act and their duration can be related to their physical properties, particularly how soluble they are in the blood. It appears that their fundamental action on the cell, too, is also physicochemical and this would explain why the chemical structures of the compounds are so diverse. The amounts needed to produce an effect are relatively large and calculations suggest that enough is present to cover a considerable portion of the surface area of cells in the brain where they are acting to produce unconsciousness. For example, the concentration of ether in the blood needed to produce anaesthesia is approximately $100\,\text{mg}\%$ (a similar concentration of ethyl alcohol produces effects on behaviour in most subjects but not unconsciousness). This is 1.35 millimoles in 100 ml and when multiplied by Avogadro's number (6×10^{23}) indicates that the number of molecules/millilitre should be about 8×10^{18}. The surface area of the cells in a gram of tissue has been estimated[5] to be about $6000\,\text{cm}^2 = 6 \times 10^{19}\,\text{Å}^2$. (The Ångstrom (Å), which is $10^{-10}\,\text{m}$, is a remarkably convenient unit for expressing molecular dimensions. The SI equivalent is 100 pm and the corresponding numbers are $6 \times 10^{23}(\text{pm})^2$ and $1 \times 10^5(\text{pm})^2$). If each ether molecule occupied only $10\,\text{Å}^2$ there would therefore be enough present in theory to cover the surface of all the cells.

The second type is characterised by finding activity only in a limited number of compounds, often having recognisable chemical features in common, which produce effects in such small doses that they can only be acting on very small parts of the cell. Many compounds, such as acetylcholine, atropine, histamine and mepyramine are active in concentrations of around 10^{-8} M, which contain about 6×10^{12} molecules/millilitre. By comparison with the calculations for ether it can be seen that these are likely to be acting only on a very small area of the cell surface.

Langley[6] used the term 'receptive substance' to describe the sites where nicotine produced its effects and Ehrlich used the word 'receptor' to describe the part of the cell which bound toxic proteins (antigens) and subsequently extended this idea to the binding of drugs.[7]

A strong reason for believing that drugs act at 'receptors' is the striking difference which is sometimes found between the activity of optical isomers. The $(+)$- and $(-)$- forms of acetyl-β-methylcholine differ about 250-fold in activity, depending on the test preparation: $(-)$- and $(+)$-hyoscyamine differ about 300-fold; $(-)$- and $(+)$-isoprenaline differ over 1000-fold in some tests. As the physico-chemical properties of the isomers, other than their effects on polarised light, are identical, their effects must be brought about by an interaction with some 3-dimensional structure.

Such a structure might be the active site of an enzyme and the word 'receptor' may be a convenient label for some undiscovered enzyme. There are many drugs, however, which are clearly not substrates for an enzyme and whose action cannot be mediated through any simple enzyme process. The effects, for example, of tetramethylammonium ions ($Me_4\overset{+}{N}$, the anion, X^-, is immaterial) might possibly be brought about by an action on some enzyme but this cannot be a simple process with the substance acting as a substrate or inhibitor.

Types of Chemical Problem

It appears, then, that some of the problems associated with the actions of drugs are physicochemical but that there are others in which there is an interaction with some three-dimensional structure, which means that the size and shape of the drug is also important. Physicochemical aspects of drug action are discussed in Chapter 2. Information about the size and shape of drugs is considered in Chapter 3 and Chapter 4 deals with the nature of processes where size and shape are important. In Chapter 5 an attempt is made to see how far all the chemical properties of a drug can be put together and related to its biological activity.

References

1. J. H. Gaddum, *Nature* (1945), no. 156, p. 463; *Pharmacol. Rev.* (1953), no. 5, p. 87.
2. J. H. Gaddum, *Pharmacol. Rev.* (1953), no. 5, p. 87; *Journal Pharm. Pharmacol.* (1953), no. 6, p. 345; *British Pharmacopoeia* (1967), Appendix XV (not in subsequent issues); *European Pharmacopoeia II* (1971), p. 441–98.
3. D. J. Finney, *Statistical Method in Biological Assay*, 2nd edn (Griffin, London, 1964); D. Colquhoun, *Lectures on Biostatistics* (Oxford University Press, 1971).
4. J. T. Litchfield and F. Wilcoxon, *Journal Pharmacol.* (1949), no. 96, p. 99.
5. A. J. Clark, *The Mode of Action of Drugs on Cells* (Arnold, London, 1933); *Handbuch der Experimentellen Pharmakologie IV* (Springer, Berlin, 1937) (in English, reprinted 1973).
6. J. N. Langley, *Journal Physiol.* (1905), no. 33, p. 374.
7. P. Ehrlich, *Münch. Med. Wschr.* (1909), p. 217.

Among the excellent general texts on statistics which should be consulted are: M. J. Moroney, *Facts from Figures*, 3rd edn (Penguin, Harmondsworth, 1964); J. K. Backhouse, *Statistics: An Introduction to Tests of Significance* (Longman, London, 1967); R. C. Campbell, *Statistics for Biologists* (Cambridge University Press, Cambridge, 1967); G. W. Snedecor and W. G. Cochran, *Statistical Methods*, 6th edn (Iowa State University Press, Ames, 1967).

TABLE 1.1: *The Distribution of t*

Degrees of freedom	Probability P		
	0.1	0.05	0.1
1	6.31	12.7	63.7
2	2.92	4.30	9.92
3	2.35	3.18	5.84
4	2.13	2.78	4.60
5	2.02	2.57	4.03
6	1.94	2.45	3.71
7	1.90	2.36	3.50
8	1.86	2.31	3.36
9	1.83	2.26	3.25
10	1.81	2.23	3.17
12	1.78	2.18	3.06
14	1.76	2.14	2.98
16	1.75	2.12	2.92
18	1.73	2.10	2.88
20	1.72	2.09	2.84
Infinite (normal distribution)	1.645	1.960	2.576

TABLE 1.2:

A *The Distribution of F (two-tailed) for P = 0.05*

Number of degrees of freedom

	in the greater variance estimate						
In the lesser variance estimate	1	2	3	4	5	10	infinite
1	648	800	864	900	922	969	1018
2	38.5	39.0	39.2	39.2	39.3	39.4	39.5
3	17.4	16.0	15.4	15.1	14.9	14.4	13.9
4	12.2	10.6	10.0	9.6	9.4	8.8	8.3
5	10.0	8.4	7.8	7.4	7.1	6.6	6.0
10	6.9	5.5	4.8	4.5	4.2	3.7	3.1
infinite	5.0	3.7	3.1	2.8	2.6	2.0	1.0

B *The Distribution of F (one-tailed) for P = 0.05 (upper) and P = 0.01 (lower)*

Number of degrees of freedom

	in the greater variance estimate						
In the lesser variance estimate	1	2	3	4	5	10	infinite
1	161	200	216	225	230	242	254
	4050	5000	5400	5600	5800	6100	6400
2	18.5	19.0	19.2	19.2	19.3	19.4	19.5
	98	99	99	99	99	99	99
3	10.1	9.6	9.3	9.1	9.0	8.8	8.5
	34	31	29	29	28	27	26
4	7.7	6.9	6.6	6.4	6.3	6.0	5.6
	21	18	17	16	16	15	14
5	6.6	5.8	5.4	5.2	5.0	4.7	4.4
	16	13	12	11	11	10	9.0
10	5.0	4.1	3.7	3.5	3.3	3.0	2.5
	10	7.6	6.6	6.0	5.6	4.8	3.9
infinite	3.8	3.0	2.6	2.4	2.2	1.8	1.0
	6.6	4.6	3.8	3.3	3.0	2.3	1.0

TABLE 1.3: *Values of the Wilcoxon U statistic. $P = 0.05$; n_1 and n_2 are the sample sizes*

n_1 \ n_2	2	3	4	5	6	7	8	9	10	12	20
2							0	0	0	1	2
3				0	1	1	2	2	3	4	8
4			0	1	2	3	4	4	5	7	13
5		0	1	2	3	5	6	7	8	11	20
6		1	2	3	5	6	8	10	11	14	27
7		1	3	5	6	8	10	12	14	18	34
8	0	2	4	6	8	10	13	15	17	22	41
9	0	2	4	7	10	12	15	17	20	26	48
10	0	3	5	8	11	14	17	20	23	29	55
12	1	4	7	11	14	18	22	26	29	37	69
20	2	8	13	20	27	34	41	48	55	69	127

TABLE 1.4: *Conversion of Percentage into Probits*

	0	1	2	3	4	5	6	7	8	9
0	—	2.67	2.95	3.12	3.25	3.36	3.45	3.52	3.60	3.66
10	3.72	3.77	3.83	3.87	3.92	3.96	4.01	4.05	4.09	4.12
20	4.16	4.19	4.23	4.26	4.29	4.33	4.36	4.39	4.42	4.45
30	4.48	4.50	4.53	4.56	4.59	4.62	4.64	4.67	4.70	4.72
40	4.75	4.77	4.80	4.82	4.85	4.87	4.90	4.93	4.95	4.98
50	5.00	5.03	5.05	5.08	5.10	5.13	5.15	5.18	5.20	5.23
60	5.24	5.28	5.31	5.33	5.36	5.39	5.41	5.44	5.47	5.50
70	5.52	5.55	5.58	5.61	5.64	5.67	5.71	5.74	5.77	5.81
80	5.84	5.88	5.92	5.95	5.99	6.04	6.08	6.13	6.18	6.23
90	6.28	6.34	6.41	6.48	6.56	6.65	6.75	6.88	7.05	7.33

CHAPTER 2

Physicochemical Problems

Rates of Reaction and Positions of Equilibrium

When considering any chemical process it is convenient to start with the Law of Mass Action, which states that the rate at which a substance reacts is proportional to its *active mass*. For gases and vapours the active mass is determined by the partial pressure which, from Avogadro's hypothesis (equal volumes of all gases contain an equal number of molecules when at the same temperature and pressure), is directly related to the concentration expressed as parts by volume. For example the partial pressure of any anaesthetic in a mixture which is 4% by volume is 0.04 times atmospheric pressure. For substances in solution the active mass is determined by the molecular concentration, if the solution behaves ideally. If the empirical formula, $C_aH_bN_cO_d$ etc. is known, the molecular weight can be calculated and the concentration is usually expressed in moles/litre, i.e. in *molarity*. Sometimes the concentration may be expressed as moles/kg solvent, i.e. in *molality*, or as the ratio of the number of molecules of solute to the total number of solute and solvent molecules present, i.e. as the *mole fraction*. If the solute and solvent do not behave ideally, and this applies particularly to salts in water, it is necessary to multiply the molarity by an *activity coefficient* to obtain the *activity*, which corresponds to the active mass.

For a reaction involving a substance A, therefore the rate of reaction $= k[A]$, where $[A]$ is the partial pressure or molar concentration (or activity) and k is a constant. For a reversible reaction,

$$A + B \rightleftharpoons C + D$$

the forward rate will equal the backward rate at equilibrium so,

$$k_{+1}[A][B] = k_{-1}[C][D]$$

where k_{+1} is the forward rate constant and k_{-1} the backward rate constant and the equilibrium constant can be written as:

$$K = \frac{k_{+1}}{k_{-1}} = \frac{[C][D]}{[A][B]}.$$

If, for example, $K = 1$, and the initial concentrations of A and B are x, the concentrations of C and D, αx, can be calculated. According to the equation 1 molecule of A and 1 molecule of B give 1 molecule of C and 1 molecule of D, so the concentrations of A and B at equilibrium will be $(1 - \alpha)x$ and

$$\frac{(\alpha x)(\alpha x)}{(1 - \alpha)x(1 - \alpha)x} = 1,$$

so $\alpha^2 = 1 - 2\alpha + \alpha^2$ and $\alpha = \frac{1}{2}$. If the original concentration of A were $2x$,

$$\frac{(\alpha x)(\alpha x)}{(2 - \alpha)x(1 - \alpha)x} = 1$$

and $\alpha = \frac{2}{3}$ so the equilibrium has been driven towards the right, increasing the concentrations of product. If both A and B were doubled, x has become $2x$ and the concentrations of products are doubled (though $\alpha = \frac{1}{2}$). If the initial concentrations of A and B were x, and if D were also present in a concentration x, at equilibrium

$$\frac{\alpha x(1 + \alpha)x}{(1 - \alpha)x(1 - \alpha)x} = 1$$

so $\alpha + \alpha^2 = 1 - 2\alpha + \alpha^2$ and $\alpha = \frac{1}{3}$ so the equilibrium has been driven to the left, depressing the formation of products. The state of the system, including the size of the equilibrium constant, is described by thermodynamics, which takes into account two parts of the reaction not shown in the equation above, the heat change and any mechanical work done (though this is only likely to be important for reactions at constant pressure in the gas phase). What happens in the reaction depends on the chemical potential energy of the molecules A, B, C and D; to understand what this form of potential energy is, it is necessary to start with the first law of thermodynamics. This states that heat and work are interconvertible.* Heat is needed to convert water at 100°C to

---

* *Calories and joules.* The calorie is a unit of heat, and is the amount needed to raise the temperature of 1 g water by 1°C, but the heat capacity of water decreases

steam at the same temperature and if the pressure is constant (1 atmosphere) some work is done, because 18 ml of water will give $\frac{373}{273} \times 22.4$ litres of steam. This work is only equivalent to a small fraction of the heat applied, however, and the molecules have acquired considerable energy, the *enthalpy* (internal energy) change for the process, $\Delta H = 37.4\,\text{kJ mol}^{-1}$; the amount of heat required is $40.5\,\text{kJ mol}^{-1}$ but the work done by expansion is $3.1\,\text{kJ mol}^{-1}$.

Energy, therefore, is stored within molecules and what happens in a reaction should depend on whether the products have less energy or more energy than the reactants. If the reactants have less energy stored in them than the products, it will be necessary to supply energy for the process to occur, whereas if the reactants have more energy stored in them the process might be expected to occur spontaneously with the evolution of heat (i.e. it will be *exothermic*). This, however, is not strictly true. There are processes, such as the solution of many salts, which proceed spontaneously, but nevertheless take in heat, i.e. they are *endothermic*.

This is because a further property needs to be considered. Not all the energy in the molecules is available. For reactions at constant volume (where no mechanical work is done by expansion or contraction) the available energy is called the *Gibbs free energy*, G, and is less than the enthalpy, H, by an amount which depends on the temperature, T, and on a property referred to as the *entropy*, S, so $G = H - TS$. The entropy, which determines the unavailable energy, can be regarded as a measure of the disorder of the system.

A reaction may therefore proceed, even though heat is taken in, because the entropy of the system increases. By convention the absorption of heat is considered to be positive, so for a reaction to occur spontaneously the change in free energy, ΔG, should be negative and the process may involve the absorption of heat provided this positive change, ΔH, is less than $T\Delta S$, where ΔS is the change in entropy. Of course the process will occur more readily if ΔH is also negative and the reaction is exothermic, but there are many processes, such as the dilution of a solution, which are mainly entropy-driven and depend on the increase in disorder

---

when it is warmed (because the structure is less organised) and it is necessary to specify the temperature. The joule is a unit of energy and at 15°C 1 calorie = 4.185 joules, whereas at 20°C 1 calorie = 4.181 joules. Although the calorie is an unsatisfactory unit, it has been convenient for experiments involving heat and has been used in the compilation of many tables of useful thermochemical and nutritional information. It seems likely that the student will need to know what a calorie is for many years to come.

brought about by the reaction. Conversely, processes which involve an increase in order (decrease in entropy) will require energy. The statement that entropy (disorder) tends to increase is one form of expressing the second law of thermodynamics.

Enthalpy, entropy and free energy are extensive properties, that is to say they depend on the amounts of material, unlike temperature, which is an intensive property and independent of amount. If G_0 is the free energy of one mole of material under standard conditions (and at constant volume), the free energy, G, of n moles of vapour with a partial pressure P is $n(G_0 + RT \ln P)$, or if it is in a concentration C dissolved in a solvent, G is $n(G_0' + RT \ln fC)$ where G_0' is the value for one mole in a molar solution at standard temperature (it is not the same as G_0 because the substance is in solution): f is the activity coefficient converting concentration to activity and is likely to be 1 if C is less than 0.01 M.

In the reaction being considered, 1 mole each of A and B produce 1 mole each of C and D. At equilibrium the partial pressures or concentrations will have adjusted themselves so that the free energy of the reactants is the same as that of the products, i.e.

$$G_{oC} + G_{oD} - G_{oA} - G_{oB} + RT \ln \frac{[C][D]}{[A][B]} = 0$$

where [A], [B], [C] and [D] are the partial pressures or concentrations if the activity coefficients $\doteqdot 1$. The change in the standard free energies, $\Delta G = G_{oC} + G_{oD} - G_{oA} - G_{oB} = -RT \ln K$, which is called the van't Hoff relation. For spontaneously occurring reactions ΔG will be negative so $\ln K$ will be positive and K will be > 1. If ΔG is positive K will be < 1 and the proportion of products will be small, though the amounts may be increased by adding more reactants.

The effect of temperature on the equilibrium constant can be calculated by differentiation:

$$\frac{d\Delta G}{dT} = -R \ln K - RT \frac{d \ln K}{dT}.$$

At constant pressure,

$$\frac{d\Delta G}{dT} = -\Delta S = \frac{\Delta G - \Delta H}{T},$$

so,

$$\Delta G - \Delta H = -RT \ln K - RT^2 \left(\frac{d \ln K}{dT}\right),$$

and,

$$\Delta H = RT^2 \left(\frac{d \ln K}{dT}\right),$$

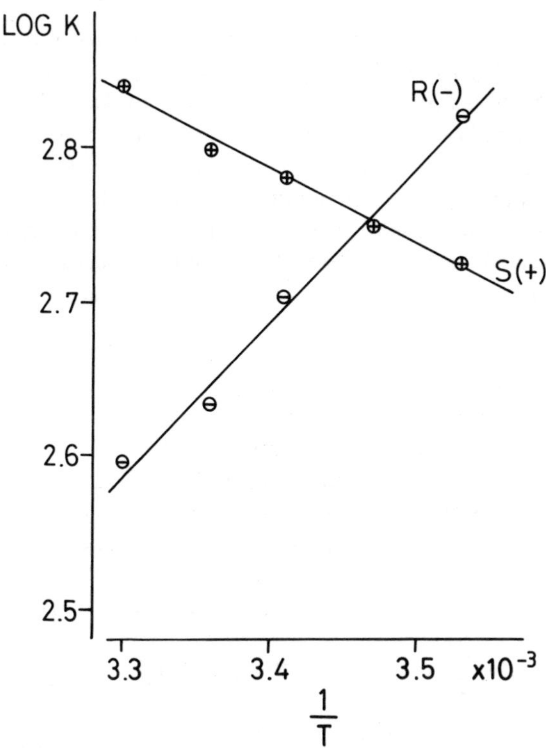

FIGURE 2.1 Effects of temperature on the affinity of $(+)$-S- ($\oplus$) and $(-)$-R- ($\ominus$) acetyl-β-methylcholine for acetylcholinesterase from red cells. The adsorption of the $(-)$-isomer is exothermic, $\Delta H = -4.6\,\mathrm{kcal\,mol^{-1}}$ $(19.2\,\mathrm{kJ\,mol^{-1}})$ and the slope is positive; lowering the temperature increases the activity. The adsorption of the $(+)$-isomer is endothermic, $\Delta H = 2.2\,\mathrm{kcal\,mol^{-1}}$ $(9.2\,\mathrm{kJ\,mol^{-1}})$. Note that the estimation of ΔH from the graph of $\log K$ against $1/T$ assumes that the effects of temperature on ΔH are small.

(From results of Belleau and Lavoie (1968) *Canad. J. Biochem.*, no. 46, p. 1397) or,

$$\left[\frac{\mathrm{d}\ln K}{\mathrm{d}(1/T)}\right] = \frac{-\Delta H}{R}.$$

The graph of $\ln K$ against $1/T$ will be a straight line with a slope of $-\Delta H/R$ (Figure 2.1), provided the change of ΔH with temperature is small. For exothermic processes, where ΔH is negative, $\ln K$ will increase with $1/T$ and therefore decrease with T. This is a thermodynamic derivation of Le Chatelier's principle, which states that systems in equilibrium alter in such a way as to annul any restraint placed upon them.

The free energy change only determines positions of equilibrium but it is possible to extend the ideas to reaction rates by assuming that the reaction proceeds through a transition state. This may be energetically unfavourable for the reactants, but favourable for the products so the rate of reaction is determined by the amount of transition state complex in equilibrium with the reactants. The function of a catalyst is to provide an intermediate complex which can be formed easily from reactants and still break down into products. Effectively it should lower the energy barrier in the way

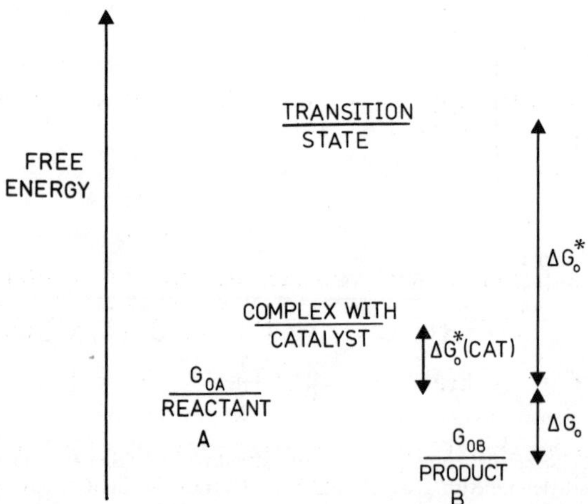

FIGURE 2.2 Although A should form B, which is more stable (ΔG_0 is negative), much energy (ΔG_0^*) is needed to form the transition state, so the reaction proceeds slowly. By adding a catalyst which can form a complex more easily than the transition state (ΔG_0^*(cat) $< \Delta G_0^*$) the reaction reaches equilibrium more quickly. The position of equilibrium should be unaffected because it depends on ΔG_0.

of the reaction (Figure 2.2), but the overall energy difference and the position of equilibrium should be unaltered.

EXAMPLE 2:1.

If a student pipettes a solution of potassium cyanide, 1 mM, and swallows 0.5 ml, what quantity does this contain (at. wt K = 39, C = 12, N = 14)?

EXAMPLE 2:2.

When preparing a solution of atropine, 10 mM, a student weighs out 73.5 mg of the sulphate ($(C_{17}H_{23}NO_3)_2$, H_2SO_4, H_2O); in what volume should this be dissolved? (At. wt C = 12, H = 1, N = 14, O = 16, S = 32.)

EXAMPLE 2:3.

A solution supposed to contain (-)-nicotine hydrogentartrate ($C_{10}H_{14}N_2$, $(C_4H_6O_6)_2$, $2H_2O$), 1 mg/ml, in fact contains this concentration of the base; what is the error?

EXAMPLE 2:4.

At 37°C, what volume of air containing HCN, 1% by volume, contains 30 mg HCN?

EXAMPLE 2:5.

The standard free energy change, ΔG_0, for the reaction

$$ATP + creatine = ADP + creatine\ phosphate$$

at 37°C is $+2\,\text{kcal mol}^{-1}$ ($8.4\,\text{kJ mol}^{-1}$). Calculate the equilibrium constant at 37°C ($R = 1.98\,\text{cal deg}^{-1}\,\text{mol}^{-1}$ or $8.314\,\text{J deg}^{-1}\,\text{mol}^{-1}$). If the concentrations of both ATP and ADP in tissues are 1 mM, calculate the ratio of the concentrations of creatine phosphate to creatine. In frog muscle the level of ATP is 3 mM, compared with 20 mM for creatine phosphate and 100 mM for creatine; calculate the level of ADP. If the standard free energy change for the reaction were zero, what would the level be?

EXAMPLE 2:6.

The affinity of (-)-hyoscine for muscarine-sensitive acetylcholine receptors in the guinea-pig ileum was estimated to be 1.74×10^9 at 37°C and 3.63×10^9 at 29°C. Calculate the free energy of adsorption at the two temperatures and estimate the enthalpy and entropy of adsorption. What does the sign of the entropy of adsorption indicate?

(From results of Barlow, Berry, Glenton, Nikolaou & Soh, *British Journal Pharmacol.* (1976) no. 58, p. 613.)

Solubility

Although the solution of a liquid or solid in a solvent is one of the simplest chemical reactions, a lack of solubility can often be most frustrating for the biologist, though sometimes it can be advantageous, as, for instance, in the use of barium sulphate in a 'meal' for the X-ray examination of the gastrointestinal tract. Barium ions produce a strong contraction of smooth muscle in concentrations of around 10^{-4}M, but the concentration of barium ions in equilibrium with a saturated solution of barium sulphate is only 10^{-5}M. A sulphate, such as sodium sulphate, is an effective antidote in the treatment of barium poisoning because it precipitates the barium ions.

A liquid solute which behaves ideally in a solvent (and obeys Raoult's law) should be infinitely soluble. There should be no heat change on solution and the process occurs spontaneously because the entropy of the system increases. There is greater disorder in the solution than in the two separate components. For a solid dissolving in a liquid, energy is needed to separate the molecules in the crystal lattice. The concentration of a saturated solution, x, expressed as a mole fraction of the solid, $n_s/(n_s+n_l)$, where n_s and n_l are the numbers of moles of solid and liquid, is related to the latent heat of fusion ΔH and the melting point, T_m, by the expression:

$$\log\frac{1}{x} = \frac{\Delta H}{2.3R}\left(\frac{1}{T} - \frac{1}{T_m}\right).$$

For example, at 25°C the solubility of naphthalene, m.p. 80°C and with a latent heat of fusion of $4.4\,\text{kcal}\,\text{mol}^{-1}$ ($18.4\,\text{kJ}\,\text{mol}^{-1}$), is

$$-\log x = \frac{4400}{2.3(1.98)}\left(\frac{1}{298} - \frac{1}{353}\right) = 0.505$$

so $x = 0.31$; this is the mole fraction in any solvent in which the substance behaves ideally.

If the solute and solvent molecules are of different sizes, however, the presence of the solute will disturb the structure of the solvent and the system will not behave ideally. For a solution in water, which is a highly structured liquid (page 183), it is most unlikely that the solute will behave ideally. Hydrocarbon groups tend to stabilise water structure and produce an increase in order and a decrease in entropy; this limits their solubility. Ions tend to break up water structure and increase entropy, thus increasing solubility, but ionised solutes are all solids whose solution depends also on the break up of the forces in the crystal which hold the ions together. As can be seen from the high melting point of ionic compounds and from their large latent heats of fusion, these forces are very strong. Considerable energy is required to separate most ions further apart than they are in the crystal lattice and in spite of the large gain in entropy to be expected from solution, solubility may be limited.

The limitations imposed by solubility on biological activity, however, depend also on the sensitivity of the receptors or enzymes which are affected. In some situations even remarkably insoluble compounds may produce effects because these are long-lasting and gradually rise to levels which are detectable. The cumulative effects of small amounts of metal ions, such as lead, have long been known and more recently examples are being discovered of similar cumulative effects of remarkably insoluble organic compounds, including many insecticides. Although their insolubility in water is usually accompanied by high solubility in structures such as insect cuticle, through which they are absorbed, they must at some point be present in an aqueous phase, but because of the length of time in which they can exert their effects their low aqueous solubility does not prevent their acting. In other types of tests where rapid responses are expected such insolubility might render the compounds inactive.

The extreme stability of particular crystal structures may be very difficult to predict and for ionised compounds only qualitative general observations can be made, such as that solubility is likely to decrease with size, as in the series Cl^-, Br^-, I^-. For non-ionised compounds where solubility is dependent largely on interactions between solute and water it should be possible to obtain some idea of the likely solubility of new compounds, provided that the solubility of some analogues has already been measured. The prediction of the likely solubility of new types of molecule however, is not really feasible.

Volume of Dilution

It would be expected that the concentration of a drug at its site of action should be related to the dose given, but the relationship will be complex because the concentration will depend on the rate of transfer to the site and the rate of removal. A useful simplification which makes it possible to set an upper limit to the concentration is provided by the idea of the *volume of dilution* of a drug. If the concentration of a drug in the bloodstream is x molar after n millimoles have been given, $x = n/V$ where V is the volume of dilution in ml. A molar solution contains 1 millimole/ml, so if 1 millimole has been given and the concentration is found to be 1 mM, the dose has been diluted 1000-fold and the volume of dilution would be 1 litre. In fact the volume of dilution is likely to be much bigger than this when a drug is given to a human adult. Some drugs appear to be dissolved in the total body fluid, which has a volume approximately 70% of the total volume (which will be about the same in litres as the weight in kilograms). Other drugs, particularly permanent ions which do not easily cross membranes, appear to be dissolved only in the extracellular fluid, which is about 25% of the total volume.

For example, if 1 mg of atropine sulphate (2.9 µmol of atropine) is given to a man weighing 73 kg ($=160$ lbs) the volume of dilution is likely to be 50 litres and the final concentration 5.8×10^{-8} M ($20 \mu g/l^{-1}$). This should produce a considerable effect. If the drug had dissolved only in the extracellular fluid ($73/4 = 18$ litres) the concentration would be expected to be $\frac{50}{18} = 2.8$ times this value.

If an antibacterial drug is active in a concentration of 10^{-6} M and it is distributed throughout the total body fluid, the dose given must be well in excess of 50×10^{-6} mol $= 0.05$ m mol. or about 15 mg for a compound with a molecular weight of 300. In practice, however, the doses usually given are about 20 times this amount because of the need to act on all the bacteria to try to prevent the emergence of resistant strains.

The calculation of likely concentrations in the blood from the volume of dilution can only be approximate. It ignores losses which may occur because some of the drug fails to be absorbed from where it has been given and because some is being removed. It can be used to set an upper limit to the concentration likely to be reached following a particular dose, or conversely to set a lower limit to the amount which must have been taken to produce a particular concentration in the blood. For example, a concentration of 80 mg% of ethyl alcohol (0.8 g/l; 17 mM) should correspond to a total intake of 40 g by a man weighing 73 kg and having a total body fluid of 50 l. This is approximately 50 ml of pure ethyl alcohol or about 120 ml of spirits. It is unlikely that less than this has been taken unless it has somehow been absorbed very rapidly, possibly if suitably diluted and on an empty stomach. The concentration in a sample of venous blood might then rise temporarily above 80 mg% because the ethyl alcohol had not had time to be distributed throughout the body but it should not stay above this concentration for long.

Rate of Removal

In most people the metabolism of ethyl alcohol after the consumption of amounts from half a pint of beer upwards appears to be independent of the concentration, i.e. it is a zero-order process. It proceeds at a rate of about $10 \, \mathrm{ml \, h^{-1}}$ ($8 \, \mathrm{g \, h^{-1}}$) and it seems as if the amounts taken saturate the enzymes converting ethyl alcohol to acetaldehyde. Half a pint (approximately 250 ml) of beer with an ethyl alcohol concentration of 5% by volume would contain $12.5 \, \mathrm{ml} = 10 \, \mathrm{g}$ EtOH so the expected blood-level in a subject weighing 73 kg would be about 0.2 g/l (4.3 mM) or 20 mg%, and 80% of this (8 g) would be removed in one hour. For a blood-level of 80 mg% in the same subject, however, corresponding to an intake of 40 g ethyl alcohol, there would be 30 g left at the end of one hour and the blood-level would be 60 mg%.

Most compounds are taken in very much smaller amounts than ethyl alcohol and it is expected that the rate of removal of the drug will be proportional to the concentration. The process is likely to involve other molecular species such as enzymes, coenzymes, oxygen or water, as well as the drug, and is almost certainly not unimolecular, but these other species are likely to be present in constant concentrations so that the reaction will appear to depend only on the concentration of drug and will be a first-order process. The rate can therefore be written: $dc/dt = -kc$, where c is the concentration of the drug, k is the *rate constant* and the negative sign indicates that the concentration decreases with time. At a time t after the start, therefore, the concentration c_t will be related to the initial concentration c_0 by the expression:

$$\ln\frac{c_t}{c_0} = -kt \quad \text{or} \quad \ln\frac{c_0}{c_t} = kt$$

so the graph of $\ln c_t$ (or $\log c_t$) against t should be a straight line with a slope of k (or $2.303\,k$). The *half-life*, $t_{1/2}$, when $c_t = c_0/2$, will be $\ln 2/k$ and $k = 0.693/t_{1/2}$. The time for the concentration to be reduced from c_0 to c_0/e is called the *time constant* and as $\ln 1/e = -1$, the time constant $= 1/k$.

TABLE 2.1: *Metabolism of Aspirin*

The plasma levels (mg %) of one subject given 1.2 g aspirin (molecular weight $= 180$) orally, were:

After	Aspirin	Total salicylic acid
30 min	1.30	5
60 min	0.80	8
90 min	0.25	6.25

(From results of Mandel, Cambosos and Smith, *Journal Pharmacol.* (1954), no. 112, p. 495.)

The process is mathematically comparable with the decay of radioactivity or with the discharge of an electric condenser. From two values of the concentration at particular times after the start of the process it is possible to calculate the rate constant, k. These values need not include the actual value at the start, c_0, which is sometimes very difficult to measure and may have to be estimated. For example, values for the concentrations of aspirin and salicylic acid in the plasma of a subject who had taken 1.2 g aspirin by mouth are given in Table 2.1. Aspirin is the acetyl ester of salicylic acid into which it is fairly rapidly hydrolysed. The concentration of aspirin declines very rapidly but the removal of salicylic acid is a much slower process with a half-life of about 24 h. The rate constant

$$k = \frac{\ln 2}{24} = \frac{0.693}{24} = 2.89 \times 10^{-2}\,h^{-1}.$$

When 90 minutes have elapsed, most of the aspirin has been hydrolysed so the concentration of salicylic acid, 6.25 mg %, should be determined almost entirely by the process for removing salicylic acid and the corresponding initial concentration of salicylic acid, c_0, would appear to be

$$c_t e^{kt} = 6.25 \times e^{0.0289 \times 1.5} = 6.25 \times e^{0.043} = 6.25 \times 1.044 = 6.53 \text{ mg} \%.$$

The molecular weights of aspirin and salicylic acid are approximately 180 and 138 respectively, so the amount of aspirin taken is equivalent to $1.2 \times \frac{138}{180} = 0.92 \text{ g}$ of salicylic acid and the volume of dilution of the salicylic acid appears to be $\frac{920}{65.3}$ litres $= 14.1$ litres. Although this figure can only be a rough guide, it suggests that the salicylic acid is probably not distributed throughout the total body fluids but only in the extracellular fluid; this is in keeping with its low pK_a (3.0; see page 64).

From the values for aspirin after one hour and one and a half hours the rate constant, half-life and initial concentration can be calculated:

$$\ln c_0 - \ln c_{t_1} = kt_1 \quad \text{and} \quad \ln c_0 - \ln c_{t_2} = kt_2,$$

so

$$\ln \frac{c_{t_2}}{c_{t_1}} = k(t_1 - t_2); \quad k = \frac{\ln (0.25/0.8)}{1 - 1.5} = 2.33 \text{ h}^{-1}$$

and the half-life $= 0.693/2.33 = 0.30 \text{ h}$ (18 min). The initial concentration, $c_0 = 0.8 \, e^{2.33 \times 1} = 8.22 \text{ mg} \%$ and the volume of dilution $= 1200/82.2 = 14.6$ litres. This again can only be a rough guide but suggests that, as with salicylic acid, the distribution of aspirin is limited to extracellular fluids and this is consistent with its pK_a (3.5). The calculated level half an hour after the dose was taken, however, is $8.22 \, e^{-2.33 \times 0.5} = 8.22 \times 0.312 = 2.56 \text{ mg} \%$, which is considerably higher than the observed value, $1.30 \text{ mg} \%$. A possible reason for this is that the absorption of the aspirin is not complete at the end of half an hour. Another result which requires explanation is the value $8.0 \text{ mg} \%$ for salicylic acid after one hour, which is bigger than the calculated initial value for salicylic acid ($6.53 \text{ mg} \%$). Among the simplifications which have been made is the neglect of possible excretion of aspirin; it has been assumed that it has been converted to salicylic acid without loss. More important, however, is the failure to take into account the rate of absorption of aspirin. Although it seems reasonable to assume absorption is complete after 1 hour, so that only the values after half an hour are affected by incomplete absorption, there is no real experimental evidence to support this.

EXAMPLE 2:7.

The concentrations of antipyrine in human plasma after the intravenous injection of 1 g are shown in the table:

Time (hrs)	Plasma concentration (µg/ml)
1	29.5
2	26.9
3	24.5
5	20.4

Calculate the half-life of antipyrine and the volume of dilution. What will the plasma concentration of antipyrine be 24 hours after the injection?

(From results of Soberman *et al.*, *Journal of Biol. Chem.* (1949). no. 179, pp. 31–42; this substance has been used for estimating total body water.)

EXAMPLE 2:8.

The plasma levels of penicillin (units/ml) following an injection of 50,000 units intravenously into a 20 kg dog are shown in the table, together with values when the animal was also given an injection of 25 mg/kg Probenecid.

Control		With Probenecid	
Time (min)	Penicillin conc.	Time (min)	Penicillin conc.
10	8.0	15	9.0
25	5.0	30	8.0
35	4.0	45	5.5
45	3.0	60	4.0
60	2.0	90	2.5
90	0.9		
120	0.4		

Calculate the half-life of penicillin in the two experiments and estimate the initial concentration and volume of dilution of penicillin. (From results of Beyer *et al.*, *American Journal Physiol.* (1951), no. 166, pp. 625–40.)

Rate of Absorption

How fast a drug is absorbed depends on how it is given as well as on what it is. The usual routes by which a drug may be administered are:

1. intravenous (*i.v.*)
2. by inhalation
3. under the tongue, also called sublingual or buccal
4. intramuscular (*i.m.*)
5. into the peritoneal cavity, intraperitoneal (*i.p.*)
6. subcutaneous (*s.c.*)
7. oral (*per os*)
8. into the rectum
9. onto the eye
10. onto the skin.

These are arranged very roughly in decreasing order of speed of absorption, though the actual rate, particularly for substances given by mouth, may vary enormously depending on what they are. Placing the drug in the blood-stream should be the quickest way of producing an effect but if the drug is a volatile liquid or a gas it could act just as quickly when inhaled, provided the actual amount which must be given is small. If large amounts are required the rate may be limited by the ventilation rate of the subject (about 6 litres/minute). For example the inhalation of a few milligrams of hydrogen cyanide will produce effects within seconds. If an anaesthetic is being administered, however, and 4 g of a substance with a molecular weight of 80 are needed to produce an effect, this is 0.05 mol or approximately $22.4 \times 0.05 \times \frac{310}{273} = 1.27$ litres of vapour. If the concentration in the inspired air is 10% and the ventilation rate is 6 l/min it would take as long as 2 mins to give enough to produce an effect.

Provided the solubility of the drug is not the limiting factor, it is the blood-flow through the area where the drug has been given which will determine the speed of absorption, except with those routes where there is a membrane between drug and blood, i.e. by mouth, under the tongue, intraperitoneally, rectally, onto the eye, and onto the skin.

The most rapidly perfused organ is the liver through which the blood flows at approximately 1.5 l/min, which is about 30% of the total cardiac output. The liver is large, however, and a better indication of the rate in relation to the size is given by the turnover time. This is the time constant for the rate of removal of the blood and indicates the time which would be needed for the system to become empty if the inflow were stopped and rate of emptying stayed constant. If the inflow is stopped, the initial rate of emptying should be kc where c is here the volume of blood in the system. If this declines exponentially c will become c/e in the time $1/k$ (page 50) and this is the *turnover time*. If the rate of removal stayed constant at kc, however, the volume c would become zero in the time $1/k$.

For the liver the turnover time is about 2 mins, for the kidneys, with a blood-flow of about 1.25 l/min but which are much smaller in size, it is about $\frac{1}{4}$ min. For the brain, with a blood-flow of about 0.75 l/min, the turnover time is about 1 min; for the heart, with a coronary blood-flow of about 0.25 l/min it is about 1 min.

In contrast the skin, muscle, skeleton and fat are poorly perfused tissues. Although the total blood-flow through the muscle and skin is not much different from that to the heart and brain the mass of tissue involved is large so there is a much longer turnover time. Values are listed in Table 2.2.

Blood-flow in the area around the tongue is high, so a few drugs which can pass across the buccal mucous membrane are absorbed rapidly when given sublingually. There is also extensive flow through the mesentery supplying the intestines so drugs which can pass from the peritoneal cavity across the mesenteric membrane act quite rapidly when given intraperitoneally; the rate is usually similar to that by the intramuscular route. Absorption can be delayed by placing the drug in an area of low blood-flow, usually by injecting it under the skin, between the skin and the muscle. Absorption can be delayed further by giving it in a form which is poorly soluble or giving it along with some solvent in which it is more soluble than in blood, e.g. penicillin G can be given as its not

TABLE 2.2: *Blood-flow through Tissues*

	Turnover time (mins)	Mass (% total)	Blood-flow (% total)
Kidneys	0.25	0.5	24
Heart	1	0.5	4
Brain	2	2	14
Liver	2	4	30
Skin	9	6	8
Muscle	35	50	15
Skeleton	65	20	3
Fat	70	17	2

Resting cardiac output $= 5.0 - 5.5$ l/min.

very soluble salt with procaine; oestradiol benzoate can be given dissolved in ethyl oleate. The usual reason for wishing to delay absorption is to try to maintain the concentration above a particular level for as long as possible and with some drugs this can be achieved by giving them orally. If the drug is not absorbed from the stomach or upper part of the gastrointestinal tract and is resistant to the enzymes present there, it may be possible to create what is effectively a depot in the lower part of the gastrointestinal tract from which the drug is gradually absorbed across the intestinal wall and into the blood-stream.

If a drug is present in a concentration C_0 in a depot which has a volume V and if the system in which it is placed has a volume V', the rate at which the concentration, C', in the system rises will depend on the rate at which C_0 falls and on the relative volumes of the depot and the system. The rate of loss from the depot will depend on the contact between the depot and the system, including such factors as the area of contact and flow rates through the region of contact (Figure 2.3). According to Fick's law it should also depend on the difference between the concentration in the depot and the system but usually the volume of the depot is very small compared with the system so any drug crossing into the system is rapidly diluted. If V is 10 ml and V' is 10 l then even when 90% of the drug has been absorbed the concentration in the system will be $0.9C_0$ (V/V'), compared with $0.1C_0$ in the depot and the ratio $C_0/C' = 0.1V'/0.9V = 111$. Even with 99% absorbed the

ABSORPTION OF A DRUG FROM A DEPOT

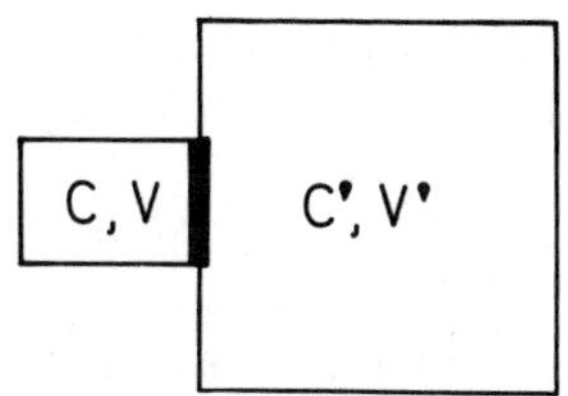

FIGURE 2.3 Absorption of a drug from a depot. C is the concentration of drug in the depot, which has a volume V, C' is the concentration in the circulation, which has an effective volume V'. The total dose of the drug $= CV + C'V'$, provided none has been lost by metabolism or excretion. The rate of absorption will depend on the circulation in the region connecting the depot with the rest of the body as well as on the concentration gradient and the ability of the drug to cross membranes in this region.

concentration in the depot should still be about 10 times that in the system.

The absorption can therefore be regarded as first-order so C, the concentration of drug in the depot $= C_0 e^{-k_a t}$, where k_a is the rate constant for absorption. The dose $D = CV + C'V'$ at any time and at the start $D = C_0 V$, so

$$C' = \frac{D - CV}{V'} = \frac{VC_0}{V'}(1 - e^{-k_a t}).$$

The rate of change of C' with time due to absorption from the depot

$$\frac{dC'}{dt} = \frac{k_a VC_0}{V'} e^{-k_a t}.$$

If the drug is being removed, however, the rate will be reduced by $k_r C'$, where k_r is the rate constant for this process, so the net change of C' with time can be written:

$$\frac{dC'}{dt} = \frac{k_a VC_0}{V'} e^{-k_a t} - k_r C'.$$

If both sides are multiplied by $e^{k_r t}$ the equation can be rearranged:

$$e^{k_r t}\frac{dC'}{dt} + k_r(e^{k_r t})C' = \frac{k_a VC_0}{V'} e^{(k_r - k_a)t}.$$

The left-hand side is the differential coefficient of $C'e^{k_r t}$ so on integrating,

$$C'e^{k_r t} = \frac{k_a VC_0}{V'} \frac{e^{(k_r - k_a)t}}{(k_r - k_a)} + \text{a constant.}$$

(The integral of e^{kx} is $e^{kx}/k + a$ constant.)

When $t = 0$, $C' = 0$, so the constant is $-k_a VC_0/V'(k_r - k_a)$ and

$$C' = \frac{k_a VC_0}{V'(k_r - k_a)}(e^{-k_a t} - e^{-k_r t}).$$

The fraction $V'C'/VC_0$ is the proportion of the initial dose (VC_0) which is present in the system. If this is called[1] P the equation becomes:

$$P = \frac{k_a}{(k_r - k_a)}(e^{-k_a t} - e^{-k_r t})$$

and if the ratio k_r/k_a is called Q, it becomes:

$$P = \frac{1}{Q-1}(e^{-k_r t/Q} - e^{-k_r t}).$$

If $k_r \ll k_a$, $Q \to 0$ and $e^{-k_r/Q} = 0$ so $P = e^{-k_r t}$.

When $k_r = k_a$, $Q = 1$ and P cannot be calculated from the above equation. The expression before integration, however, becomes:

$$e^{kt}\frac{dC'}{dt} + k\,e^{kt}C' = \frac{kVC_0}{V'}$$

(where $k = k_a = k_r$) and the integral becomes $C'e^{kt} = kVC_0 t/V' + a$ constant. As $C' = 0$ when $t = 0$, the constant $= 0$ so the equation becomes

$$C' = \frac{kVC_0 t}{V'}e^{-kt} \quad \text{and} \quad P = kt\,e^{-kt}.$$

It is convenient to plot P against $k_r t$ (Figure 2.4).

Absorption will be maximal when the rate of absorption,

$$\frac{dP}{dt} = \frac{k_a}{(k_r - k_a)}(-k_a e^{-k_a t} + k_r e^{-k_r t}) = 0$$

so,

$$k_a e^{-k_a t} = k_r e^{-k_r t}, \quad \text{or} \quad \ln k_a - k_a t = \ln k_r - k_r t$$

and the time when absorption is maximum,

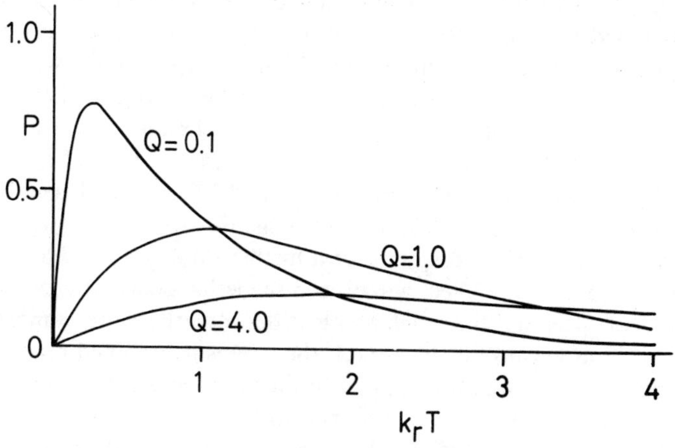

FIGURE 2.4 The proportion, P, of the original dose which is present plotted against the time multiplied by the rate constant for removal, i.e. $k_r t$, for various values of Q, the ratio of the rate constant for removal to that for absorption, i.e. k_r/k_a.

$$t_{\max} = \frac{1}{(k_a - k_r)} \ln\left(\frac{k_a}{k_r}\right)$$

and

$$k_r t_{\max} = \frac{Q}{1-Q} \ln\left(\frac{1}{Q}\right).$$

To calculate the corresponding value of P it is necessary to rewrite the equation in the form:

$$P = \frac{k_a e^{-k_r t}}{(k_r - k_a)}(e^{-(k_a - k_r)t} - 1)$$

and as:

$$e^{t_{\max}} = \left(\frac{k_a}{k_r}\right)^{[1/(k_a - k_r)]}$$

$$P_{\max} = \left(\frac{k_a}{k_r - k_a}\right)\left(\frac{k_a}{k_r}\right)^{[-k_r/(k_a - k_r)]}\left[\left(\frac{k_a}{k_r}\right)^{-1} - 1\right] = \left(\frac{k_a}{k_r}\right)^{[k_r/(k_r - k_a)]}$$

When $k_a = k_r = k$,

$$\frac{dP}{dt} = k e^{-kt} - k^2 t e^{-kt}$$

so when $dP/dt = 0$, $kt = 1$ and $P = e^{-1} = 0.368$ (with $t_{\max} = 1/k$).

These calculations may be used, for instance, to obtain some idea of the relation between dose and an effect such as sleeping time, if it is assumed that the effect lasts for as long as the concentration is greater than a lower limit but less than an upper limit which is lethal. From Figure 2.4 it can be seen that for a drug with $Q = 0.1$ a dose, x, will be present in an amount $>0.1x$ for approximately the time $= 2.4/k_r$. If the dose is halved, the amount present must be $>0.2(x/2)$ which lasts for approximately $1.7/k_r$; for a dose $= x/4$, $P = 0.4$, which lasts for $0.95/k_r$, but if the dose is $x/8$, $t = 0$. If the dose is doubled, $t = 3.0/k_r$ and for a further doubling $t = 3.6/k_r$ but somewhere the peak effect will exceed the lethal concentration. The graph of sleeping time against log dose is therefore almost a straight line, because with $Q = 0.1$, the time is determined primarily by the rate of excretion and it is only with the lower doses that the rate of absorption contributes significantly.

The expressions may also be applied to the situation where a drug is converted into an active form in the body. As an approximation, the rate of formation of the active form can be regarded as being equivalent to the absorption process though it may be necessary to consider also the rate of absorption of the inactive precursor. With three or more exponential processes linked together the mathematics are complex but may be solved with help from a computer.

The theory is an improvement on the simple use of volumes of dilution as it takes into account the time-course of events. Because of the different blood-flow through the various organs, however, it is too simple to give an accurate interpretation of the early phases of drug absorption and distribution when the drug is placed in an area which is rapidly perfused (page 54). A common example is the intravenous injection of an anaesthetic such as thiopentone sodium. A dose of 0.5 g (1.9 mmol) injected in a small volume (about 2 ml; the solution is about 1 M) into a vein in the arm is likely to produce anaesthesia in an adult lasting for perhaps 10 min. If this were dissolved in the total body fluid (50 l) the expected concentration would be about 40 µM but the concentration reaching the brain immediately after the injection is likely to be more than 10 times this. The drug becomes rapidly redistributed, however, and as it reaches areas which are more slowly perfused the concentration in the brain falls and anaesthesia wears off, even though the drug has not been eliminated to any appreciable extent. This particular drug has a very high solubility in fat, with a partition coefficient between fat and water of over 500:1.[2] As the fat constitutes between 15 and 20% of the total body weight, this will effectively increase the volume of dilution between 75 and 100-fold but the dilution can only occur when the circulation has had time to produce some sort of balance in the distribution of the drug in the various parts of the body, including those areas which are more slowly perfused.

EXAMPLE 2:9.

The half-life of a drug is 8 h and the half-time for oral uptake is 0.5 h. How long will it take before 25% of the initial dose has been absorbed and for how long will absorption stay above this level? (Construct a graph of P against time or $k_r t$.) When will peak absorption occur and what proportion of the initial dose will be absorbed at this time? If the maximum level which can be tolerated is 4 times that needed therapeutically, will the subject survive a second dose of the same size given after 8 h?

EXAMPLE 2:10.

The table below shows the plasma and brain concentrations (mg/kg) of thiopentone in a dog which had received an intravenous injection of 40 mg/kg:

Time (min)	Plasma	Brain
10	48	72
20	27	
40	19	37
65	16	
85	14	26
150	12	17

Use the first two results for plasma and for brain to calculate the initial volume of dilution (per kilogram body weight) and compare the values with the volume calculated from the last two results. What may the difference indicate?

(From results of Brodie and Hogben, *Journal Pharm. Pharmacol.* (1957), no. 9, p. 345.)

Partition Coefficients and Crossing Membranes

Because the rates of circulation are different it is useful to think of the body as a series of compartments connected by pipes of different diameter (Figure 2.5) but it is also clear that there are barriers which limit the transport of some drugs to some compartments. Some of these arise because the drug must cross a membrane in order to reach the compartment. Although there are transport systems in the cell membrane which carry substances into and out of the cell, these are usually active processes requiring energy which are specific for amino acids, glucose, or other substances which have a particular part to play in the life and function of the cell. The cell membrane is a protein and lipid barrier and substances which have appreciable solubility in water and in lipid may diffuse through the membrane passively (see below). It is usual to find, therefore, that organic compounds which are charged do not cross membranes. Quaternary ammonium salts such as tetramethylammonium, $Me_4\overset{+}{N}$ (the anion is immaterial), for example, are very poorly absorbed when taken by mouth. On the other hand, substances which are not charged are often

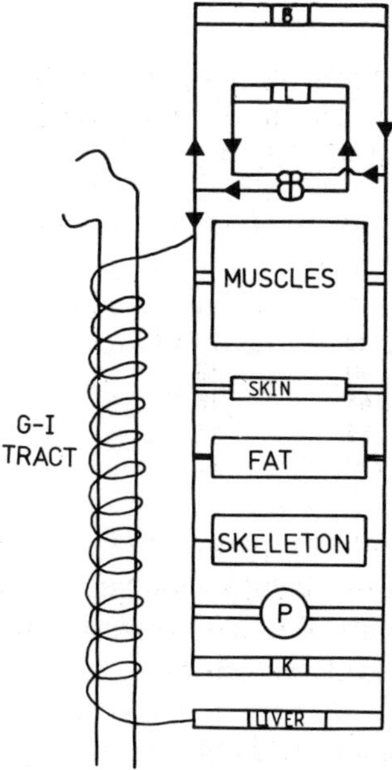

FIGURE 2.5 The body represented as a series of compartments connected by pipes of different size. The compartments are drawn in proportion to their mass and the inlets and outlets are in proportion to the blood-flow (Table 2.2). B represents the brain, L represents the lungs, K represents the kidney, and P represents the plasma proteins which are important in the binding of some drugs but not others, so this is shown as a circle whose size has no significance. G-I tract is the gastrointestinal tract.

sufficiently lipid soluble to be able to cross membranes passively quite readily, e.g. ethyl alcohol.

The ability to cross membranes is obviously important in determining the access of a drug to sites beyond a membrane, for example through the cornea, and it is also important in determining the absorption of a drug when given by mouth, because it makes it possible for a drug to diffuse from the lumen of the gastrointestinal tract through the wall and into the bloodstream. Ability to cross membranes is also important in determining the retention of a drug within the circulation. When blood with the drug dissolved in it enters the glomerulus of the kidney most drug molecules have a molecular weight less than 20,000 and are small enough to pass into the tubular filtrate, so they will be excreted unless reabsorbed during their passage along the tubule. There are also active processes, usually in the distal portion of the tubule, which excrete some materials into the tubular fluid. Reabsorption can only occur if the drug can pass across the tubular wall and back into the blood-stream. Drugs which can do this will be retained and removal of the drug will depend on its conversion by enzymes into species which are not able to cross membranes. Usually this involves the production of metabolites which are charged, e.g. products linked to glucuronic acid. Ability to cross membranes accordingly largely determines the removal of a drug as well as access to its site of action.

According to *Fick's law of diffusion* the rate of transport of material

$$\frac{\mathrm{d}x}{\mathrm{d}t} = -DA\frac{\mathrm{d}C}{\mathrm{d}L}$$

where $\mathrm{d}C/\mathrm{d}L$ is the concentration gradient, A is the cross-sectional area across which diffusion is taking place and D is a constant, called the diffusion coefficient. According to *Graham's law*, the rate is also proportional to $M^{1/2}$ where M is the molecular weight, so $DM^{1/2}$ should be constant. For diffusion through a membrane the diffusion coefficient D must be replaced by another constant, Z, the permeability coefficient. If the membrane were simply a lipid barrier it might be expected that $ZM^{1/2}$ would be a constant (but if it is porous the pore-size may enter into the permeability and the relation would not hold for large molecules). It is found experimentally that permeability in many systems is associated with a high oil-water partition coefficient.[3] The relationship between the permeability coefficient and the partition coefficient is complex but if the process is relatively rapid, $ZM^{1/2}$ is approximately proportional to the partition coefficient (at a constant temperature).

The partition coefficient, P, is therefore extremely important because it determines the distribution of the drug throughout the body and the rates at which it becomes distributed across membranes, which affect access and removal of the drug. To calculate the partition coefficient it is necessary to allow the distribution of the drug between the two phases to reach equilibrium and to measure the concentrations in each phase. It may be necessary to shake for some time to achieve equilibrium and to allow a long period for the two phases to separate. This process may be speeded up by centrifugation which should assist the return to the heavier layer of droplets dispersed among the lighter layer. With aromatic compounds and other substances which absorb visible or ultra-violet light the concentrations may often conveniently be estimated with a spectrophotometer. This may involve diluting the solutions to obtain a concentration which can be measured accurately, and therefore assuming that the absorption, $E, = \log(I_0/I) = \varepsilon \times C \times l$ where I and I_0 are the intensity of the light after passage through the solution and solvent respectively, C is the concentration of solute, l is the length of the light path and ε is a constant, usually called the molar extinction coefficient. Standard solutions must be prepared for calibration and if a range of these is chosen, it is possible to check for compliance with Beer's law. Complications will arise if one of the solvents itself absorbs to any extent but it may be possible to avoid making a measurement with this phase by taking a known amount of material initially and estimating the amount in one phase by subtraction from that in the other. This is not as accurate as making a direct comparison because it may bias the results. If the concentration in one phase has been underestimated, that in the second phase calculated by subtraction will be overestimated and the error in the ratio will be large.

For example, 39.50 mg phenol (m wt 94.11) was shaken with 30 ml n-octanol and 300 ml water for 3 hrs. The optical density of the aqueous layer, after centrifugation, indicated a concentration of 3.28×10^{-4}M phenol calculated from the absorbance of standard solutions of phenol in water. The octanol layer was centrifuged and diluted 1 to 25 with octanol and the optical density indicated a concentration of 3.88×10^{-4}M phenol (calculated from standard solutions of phenol in octanol). From these two concentrations the partition coefficient

$$P = 25 \times \frac{3.88}{3.28} = 29.6.$$

The amount of phenol taken is

$$\frac{39.5}{94.11} = 0.4198 \text{ mmol}$$

and the amount in the water layer is $300 \times 0.328 \times 10^{-3} = 0.0984$ mmol so the octanol layer should contain 0.3214 mmol in 30 ml and the concentration is 1.071×10^{-2} M. Accordingly

$$P = \frac{10.71}{0.328} = 32.7.$$

The value of P calculated from the weight of phenol and the concentration in the octanol layer, however, is 22.6.

The partition coefficient depends not only on the compound but on the nature of the liquids between which the compound is being distributed. One of these is likely to be water and the other to be a substance with which water is not miscible to any great extent. Such substances may be described as lipid-like but do not behave identically so values of P will vary from one system to another. The possibility of relating values obtained with one lipid-like phase to those obtained with another is discussed later (page 74).

Ionisation and pK_a

Many drugs are acids, e.g. barbiturates, aspirin, sulphonamides, penicillins, or bases, e.g. local anaesthetics, antihistamines, antimalarial drugs, and so exist as an equilibrium mixture of ionised and unionised species. Though the fundamental action of the drug may depend upon the ionised species, it is the unionised form which is likely to have some degree of lipid solubility and so be able to cross membranes. The biological activity of these compounds should therefore be determined to a very large extent by the position of the equilibrium between ionised and unionised forms. For an acid the ionisation can be written:

$$HA \rightleftharpoons H^+ + A^-$$

and the *equilibrium constant*

$$K_a = \frac{[H^+][A^-]}{[HA]}.$$

For acetic acid the proportion of ion present is small; in water at $25°C$ K_a is 1.7×10^{-5}. It is convenient to take the logarithm of the reciprocal of this value, pK_a which is 4.8. For acids which dissociate to a smaller extent, K will be smaller still and pK_a will be larger. For acids which dissociate to a greater extent pK_a will be smaller; it will be log 1 $(=0)$ when $[H^+][A^-] = [HA]$ and negative when $[H^+][A^-] > [HA]$. For a base the ionisation can be written:

$$B + H_2O \rightleftharpoons BH^+ + OH^-$$

and the equilibrium constant

$$K_b = \frac{[BH^+][OH^-]}{[B]}$$

(H_2O is constantly in excess). For ammonia the proportion of ions present is small; in water at $25°C$ K_b is 1.6×10^{-5} and pK_b is 4.8.

It is more convenient to consider the ionisation process of the conjugate acid of the base. however, which can be written:

$$BH^+ \rightleftharpoons B + H^+$$

(the concentration of the anion, X^-, is constant) so

$$K_a = \frac{[B][H^+]}{[BH^+]} = \frac{K_w}{K_b}$$

where K_w is the ionic product of water, $[H^+][OH^-]$, and $pK_a = 14.0 - pK_b = 9.2$ (at $25°C \log K_w = -13.996$).

One advantage of using pK_a values for both acids and bases is that strong acids have small values and strong bases have large values which produces a convenient distinction between them. Another reason for using pK_a values of bases is that the actual measurements are nearly always made with solutions of salts of the base, rather than with the base itself. This should be particularly true with drugs which are nearly always marketed as salts because these are usually stable crystalline solids, whereas the bases are often liquids or unstable and sometimes much more difficult to dissolve in water.

From the expression

$$K_a = \frac{[H^+][A^-]}{[HA]}, \quad pK_a = pH + \log\frac{[HA]}{[A]} = pH + \log\left(\frac{\text{unionised}}{\text{ionised}}\right)$$

and for the conjugate acid of a base

$$K_a = \frac{[B][H^+]}{[BH^+]}, \quad pK_a = pH + \log\frac{[BH^+]}{[B]} = pH + \log\left(\frac{\text{ionised}}{\text{unionised}}\right).$$

The proportion of material ionised can be altered by adding alkali which will increase the concentration of A^- ions and decrease the concentration of BH^+ ions. When half the amount of alkali needed to react with the acid has been added, $[HA] = [A^-]$ and $[BH^+] = [B]$ so the pH of this half-neutralised solution will equal the pK_a. The simplest method of measuring the pK_a, therefore, is to add alkali to a solution of the acid, or to a solution of the conjugate acid of a base, and to measure the pH with, for example, a glass electrode. Usually the pH is measured after several additions of alkali, rather than only after adding half an equivalent. From the amount of material taken and the amount of alkali added the ratio of unionised to ionised species can be calculated and the pK_a calculated by adding the log of this ratio to the pH for an acid, and subtracting it from the pH for the conjugate acid of a base. A series of estimates of pK_a is thus obtained and these should be constant. The amount of alkali added must not exceed one equivalent and preferably produce a ratio of unionised to ionised between 0.1 and 10.0. It is also necessary to control the temperature and to protect the reaction from atmospheric carbon dioxide.

Because strong solutions of ions do not behave ideally their concentrations should be converted into 'activities' by multiplying by an activity coefficient, f. The glass electrode actually measures hydrogen ion activities,* not concentrations, so the 'mixed' pK_a obtained in the potentiometric titration can be converted into 'true' or 'thermodynamic' pK_a by subtracting $\log f$ for the titration of an acid, HA, or adding it for the titration of the conjugate acid of a base BH^+. Because f is less than 1, $\log f$ is negative, so the true pK_a of an acid is bigger than its mixed pK_a whereas that of a base is smaller. For dilute solutions, however, $f \to 1$ and it is 0.9 for a concentration of $10\,mM$ so the difference between the two pK_a values is less than 0.05.[4]

With some substances the ionised and unionised species have different absorption spectra. With indicators this difference can be seen by eye and for many other compounds there is a difference in the ultra-violet region, e.g. for compounds where the ionising group is attached to an unsaturated system such as a benzene ring. If the compound is made up in buffer at different pH values which are

---

*The glass electrode consists of a silver wire covered with silver chloride in hydrochloric acid enclosed in a thin bulb of suitable glass, which acts as a membrane and allows chemical continuity with the hydrogen ions in the solution in which the glass bulb is immersed. The potential across the membrane is determined by the free energy needed to transfer hydrogen ions: $E = -\Delta G/nF$ where F is the electrochemical equivalent (the Faraday) so nF represents the total charge carried. The potential E is therefore $(RT/nF)\ln a_1/a_2$, where a_1 and a_2 are the hydrogen ion activities on either side of the glass. The cell is completed by a second electrode, usually a calomel electrode, which is often incorporated into the same assembly. It is calibrated with standard buffers but note that the ionisation of the buffers depends on the temperature as well as the potential from the cell. For instance, the pH of a 0.01 molal solution of sodium tetraborate (borax) is 9.18 at 25°C and 9.08 at 37°C.

known, the ratio of the ionised to the unionised species may be calculated from the spectra and the pK_a calculated.

Once the pK_a of a substance is known it is possible to make the above calculations in reverse and calculate the proportion of ionised to unionised species for any pH. If the pH is more than 2 log units less than or greater than the pK_a the ratio of one species to the other will be greater than 100:1 and if, for example, the ionised form predominates, the proportion of unionised form is unlikely to be sufficient for passive transport across membranes. The ratio is particularly sensitive to changes in pH when it is close to the pK_a, however, and a change of pH from $pK_a - 0.3$ to $pK_a + 0.3$ will change the ratio of $[HA]:[A^-]$ or $[BH^+]:[B]$ from 2:1 to 1:2.

Although the buffering capacity of the blood will prevent such big changes occurring in the circulation, the range of pH in the gastrointestinal tract is very large, from nearly 1 in the stomach to about 8 in the lower parts of the tract. For many drugs with a pK_a between 1 and about 9 there is therefore likely to be an appreciable proportion of either ionised or unionised species somewhere in the tract. The rate of absorption, however, will depend upon how far down the drug must go for the unionised form to be present in amounts sufficient for absorption. Strong bases, for example, will have to travel a long way before the pH is alkaline enough for absorption, whereas acids may be absorbed much more rapidly from higher up, possibly even from the stomach.

A range of pH, though nothing like as big, can also occur in the tubular filtrate. This may be made acid by giving ammonium chloride, because the ammonium ions are metabolised into urea which is excreted and there is a net gain of HCl. It may be made alkaline by giving sodium bicarbonate or the sodium salts of organic acids such as sodium acetate or sodium citrate, whose anions are metabolised to carbon dioxide with a net gain of sodium carbonate. The changes in pH which can be produced range from about pH 5 to pH 9 and it is possible to affect the excretion of drugs with a suitable pK_a. A more acid urine will increase the excretion of bases and a more alkaline urine will increase that of acids. This can be made use of, for example, in order to shorten or prolong the action of a dose of amphetamine ($pK_a = 9.9$). Changes in pH can also be used by the pharmacologist in order to try to find out whether the ionised or unionised form of a drug is the active species by comparing the activity at different pH values relative to some other drug, often a quaternary ammonium salt, whose ionisation is not altered by the pH change.

For example, nicotine has two basic groups with pK_a values of 3.1 and 8.0 at 25°C. If choline phenyl ether is 1.25 times as active as nicotine on the isolated frog rectus muscle at pH 7.6, and only the singly protonated nicotinium ion is biologically active, how will the activity alter as the pH of the solution bathing the preparation is changed to 7.7, 7.8 and 8.0? At pH 7.6 the ionisation of the weaker basic group (pK_a 3.1) can be neglected and the ratio of singly protonated nicotine to the unchanged base will be $10^{pK_a - pH}:1 = 2.5:1$. The proportion of nicotinium ion is therefore 2.5/3.5 and the activity of choline phenyl ether relative to the nicotinium ion is $1.26 \times (2.5/3.5) = 0.89$. At pH 7.7 the ratio of ionised to unionised is $2:1$ so the relative activity of choline phenyl ether will be $(3/2) \times 0.89 = 1.33$; at pH 7.8 it will be $(2.6/1.6) \times 0.89 = 1.45$ and at pH 8.0 it will be $2 \times 0.89 = 1.78$.

Most measurements of pK_a are made at 25°C. The values at 37°C will be different and in some situations the difference can be important. The dissociation of organic bases is much more affected by temperature than that of organic acids and appears to be roughly proportional to the pK_a. For strong bases such as ethylamine (pK_a 10.7) it is about 0.03 units/deg C whereas for aniline (pK_a 4.6) it is about 0.015 units/deg C. At 37°C, therefore, the pK_a of the stronger basic group in nicotine is likely to be $8.0 - 12 \times 0.025 = 7.7$. The pH range 7.6 to 8.0 therefore produces a change from the cation being the major species to the base predominating. The ratios of ionised to unionised are $1.26:1$ at pH 7.6 and $0.5:1$ at pH 8.0. If the activity of the choline phenyl ether relative to the nicotinium ion remains 0.89, the observed activity should be 1.59 at pH 7.6 and 2.7 at pH 8.0. The effect on activity of altering pH is proportionately greater because the pK_a lies nearer the middle of the pH range being tested, rather than at one end (8.0).

EXAMPLE 2:11.

The pK_a of arecoline at 35°C is 7.6. Burgen, *British Journal Pharmacol.* (1965), no. 25, p. 4, has calculated that the ion is 1.42 times as active as carbachol on the guinea-pig ileum and that the unionised form is inactive. Calculate the activity of arecoline relative to carbachol at pH 6.6, 7.0, 7.6, 8.0 and 8.6.

EXAMPLE 2:12.

In a human weighing 70 kg the total body fluid is about 50 l. If 50 mg (0.12 mmol; 1 tablet) of mepyramine is taken by mouth and the pK_a of mepyramine is 8.6, the pH of the stomach is 1.6 and the volume of fluid in it is 200 ml, calculate the approximate concentration of mepyramine in the body fluid at pH 7.6. Assume that the unionised form of the drug can diffuse freely through the stomach wall. Repeat the calculations of the concentration in body fluid at pH 7.6, assuming that the contents of the stomach have passed into the duodenum and the pH is 6.6 and again when they have reached the ileum the pH is 7.6. Assume that there is no significant metabolism of the drug and neglect the amounts absorbed and any volume changes in the upper part of the tract, i.e. assume that you have 50 mg dissolved in 200 ml at pH 6.6 or pH 7.6.

EXAMPLE 2:13.

The pK_a of a basic local anaesthetic is 7.5. If a concentration of 10^{-3} g ions/l in the neighbourhood of the fibres is required to block conduction, calculate the total concentration of drug required in the fluid bathing a sheathed nerve trunk in order to block conduction when the pH of the fluid is (a) 6.5, (b) 7.5 and (c) 8.5, if two different assumptions are made:

 (i) that the pH of the fluid enclosed by the sheath is 7.5 and is not affected by the external pH;
 (ii) that the pH within the sheath is the same as the external pH.

Zwitterions. A substance which contains both an acidic and a basic group can form an ion which carries both positive and negative charges and is called a *zwitterion*. Amino acids are a common example. The pK_a values for acetic acid and ammonia are 4.8 and 9.2, respectively, so at pH values lying within this range it would be expected that amino-acids would exist almost exclusively as zwitterions, with the carboxyl group in the form $-COO^-$ and the amino group in the form $-\overset{+}{N}H_3$. This is true for α-amino acids and also for aliphatic amino acids such as γ-aminobutyric acid but for aminobenzoic acids the problem is more complex. The pK_a of benzoic acid (4.2) is close to that of aniline (4.6). There is still appreciable zwitterion formation, however, because the charged $-\overset{+}{N}H_3$ group increases the strength of the carboxyl group. For *m-*

aminobenzoic acid, for example, the zwitterion constant,

$$\frac{[H_3\overset{+}{N}-C_6H_4COO^-]}{[H_2N-C_6H_4COOH]}$$

has been estimated to be just under 3.[4]

The pK_a of phenol (9.9) is slightly greater than that of phenethylamine (9.8) and it might be expected that the $-\overset{+}{N}H_3$ group would lose its proton before the $-OH$ group and no zwitterions would be formed. The charged $-\overset{+}{N}H_3$ group, however, again increases the strength of the $-OH$ group and in tyramine the zwitterion constant appears to be about 3. In dopamine,

HO—⬡—$CH_2\,CH_2\,NH_2$

HO

the second phenolic group makes the substance a stronger acid still (the pK_a of the trimethylammonium salt of dopamine is 8.7) and the zwitterion constant for dopamine is about 6. With noradrenaline, however, the alcoholic hydroxyl group is base-weakening. The pK_a of $PhCHOHCH_2NH_2$, for instance, is 8.9, almost one unit less than that of phenethylamine, and the zwitterion constant is between 1 and 2.[5] With 5-hydroxytryptamine the zwitterion constant appears to be about 1.

With all these phenolic amines the first pK_a is considerably greater than 7.6 and the predominant species at biological pH has the phenolic group intact and the amino group positively charged. With noradrenaline and adrenaline, for example, pK_1 is 8.5 so the proportion of zwitterion present at pH 7.6 is small (about 5%). With other phenolic amines pK_1 is larger (8.8 for dopamine, 9.2 for tyramine) and the proportion of zwitterion present at pH 7.6 is no bigger, because the increase in pK_1 offsets any increase in zwitterion constant.

It is not known whether zwitterions are of importance for the biological actions of these phenolic amines, which seem rather to depend on the presence of intact $-OH$ groups, but it is quite clear that there must be big differences between receptors for phenolic amines and catecholamines and receptors for amino acids, which are almost exclusively zwitterions.

Relations between Reaction Rates and Equilibrium Constants

The transport, distribution, and removal of a drug depend to a large extent on physical properties which determine rates of movement of the drug and the position of equilibria, such as the partition of a drug between two phases and the extent of ionisation. In theory these chemical properties should be related to the molecular properties of the drug, particularly its size and the distribution of electrons. Although it may be difficult to consider electron distribution throughout the molecule as a whole, it is often possible to assess changes in distribution brought about by changing the chemical structure and Hammett[6] has shown how these can be used quantitatively.

The effect of a group on electron distribution is most obvious from the change it produces in the pK_a of an acid or base. An electron-withdrawing group should increase the ionisation of an acid by weakening the bond to hydrogen and this will reduce the pK_a. For example, the pK_a of chloracetic acid at 25°C is 2.81, compared with 4.76 for acetic acid. Electron-releasing groups should strengthen the bond to hydrogen and so produce a weaker acid; the pK_a of propionic acid is 4.88. The same will apply, qualitatively, to the bond between nitrogen and hydrogen in the conjugate acid of a base, but weakening the bond and lowering the pK_a will make the substance a weaker base and strengthening it will make it a stronger base. For example, ethylamine, pK_a 10.70, is a slightly stronger base than methylamine, pK_a 10.65. The extent of the effect on pK_a will depend on how far the substituent is from the ionising group and the nature of the intervening systems. Where there are systems containing π-electrons the effects can be relayed in a remarkable way but with saturated systems the effects are not passed on to any great extent.

It might be expected that substituents which weakened the link to hydrogen in an acid might also weaken the link to methyl in its methyl ester so the hydrolysis might proceed faster. Similarly, the rate at which an amino group could react with methyl iodide might be related to the strength of its link to a proton; substituents which weakened the bond to the proton, making it a weaker base, should also make it more difficult to react with methyl iodide. Hammett pointed out that these qualitative ideas could be applied quantitatively to a very large number of processes if the logarithms of rate constants and logarithms of equilibrium constants are plotted against each other.

Figure 2.6 shows values of the logarithms of the rate constants for the alkaline hydrolysis of ethyl esters of substituted benzoic

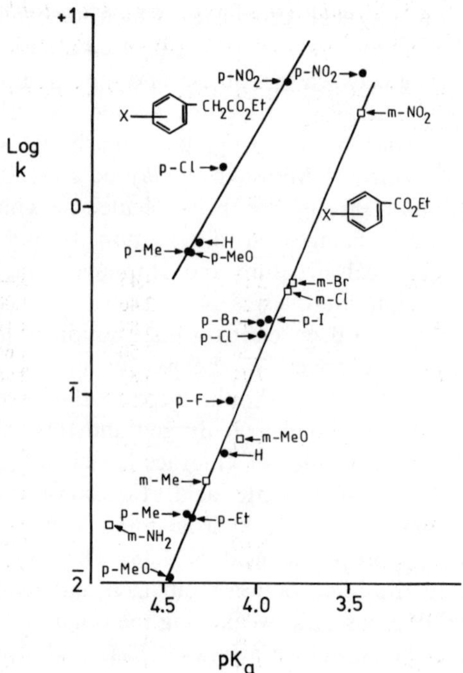

FIGURE 2.6 The logarithms of the rate constants for the alkaline hydrolysis of ethyl esters of benzoic acid (lower graph) and phenylacetic acid (upper graph) are plotted against the pK_a values of the acids, with the scale reversed so that acid-strengthening groups appear to the right (as they would do if values of σ were plotted instead). Note that the effects of substituents on both the rate constants and the ionisation constants is much smaller in the derivatives of phenylacetic acid than in those of benzoic acid.

(Redrawn from results collected by Hammett, *Chem. Review* (1935), no. 17, p. 131.)

acids plotted against the pK_a of the corresponding acid. The upper line shows the graph for ethyl esters of substituted phenylacetic acids and their corresponding acids. The pK_a scale has been reversed, so that it indicates increasing electron-withdrawing character. The *m*-nitro group, for instance, increases the ionisation of benzoic acid and the rate of hydrolysis of the ethyl ester is fast. The electron-releasing *p*-methoxy group decreases the ionisation of benzoic acid and slows down the hydrolysis of its ethyl ester. A similar situation is seen with the phenylacetic acid esters, though the effects of substituents on rate and on ionisation are less marked, because the ring is separated from the carboxyl group by a methylene group.

TABLE 2.3: *Prediction of* pK_a, *based on tables of* pK_a (Kortum, Vogel and Andrussow; Perrin).

The pK_a values of substituted benzoic acids and anilines at 25°C are:

	Benzoic acid	4.17	
	Aniline	4.60	

	o	m	p
NO_2	2.17	3.45	3.44
	−0.28	2.46	1.00
F	3.27	3.87	4.14
	3.20	3.50	4.65
Cl	2.94	3.83	3.98
	2.64	3.52	4.00
Br	2.85	3.81	3.97
	2.53	3.55	3.88
I	2.86	3.85	3.93
	2.60	3.60	3.78
Me	3.91	4.27	4.37
	4.42	4.70	5.08
MeO	4.09	4.09	4.47
	4.52	4.23	5.34
HO	2.98	4.08	4.58
	4.70	4.25	5.55
H_2N	4.98	4.79	4.92
	4.80	5.00	6.20

σ *Values*

The experimental results fit a straight line quite convincingly, as do hundreds of other examples which have been examined. If a straight-line relationship is to be expected, it will be possible to use it predictively. When results are known for two compounds the line can be drawn and the log rate constant can be estimated for any compound whose pK_a is known. It is convenient to have a standard system to which others can be related and the ionisation of substituted benzoic acids in water at 25°C is an obvious choice. To avoid negative signs, the effect of a substituent, σ, is defined as $pK_{a0} - pK_a$, where pK_{a0} is that of benzoic acid itself. For a m-nitro group, therefore, $\sigma = 4.17 - 3.45 = 0.72$; σ is in fact $\log(K/K_0)$ for the ionisation of benzoic acids. Values of σ can be calculated from values of pK_a such as those in Table 2.3.[7] For a reaction which fits a Hammett relation, such as the hydrolysis of ethyl benzoates, the effect of a substituent on $\log K$ should be directly proportional to σ, its effects on the pK_a of benzoic acid. Accordingly $\log(k/k_0) = \sigma\rho$, where k and k_0 are the rate constants for the substituted and unsubstituted esters and ρ is the proportionality constant. Once two estimates of rate constant have been made for compounds whose σ value are known, it is possible to calculate the reaction constant ρ and hence values of the rate constant for any other compound whose σ value is known.

Extension to Partitioning; π Values

In most of the original work, Hammett was concerned with correlating the logarithms of rate constants with σ values but the ideas should clearly be applicable to correlating equilibrium constants, because these are derived from the rate constants for the forward and backward reactions. An obvious extension would be to partition coefficients. These are likely to be determined by other factors, including size, as well as electron distribution, and it may be too optimistic to expect that they will be related only to σ values. It is highly likely, however, that there is a good correlation between partition coefficients obtained in different solvent systems, so although the actual partitioning processes which occur in a biological system may not be known, it may be possible to obtain information about the effects of changes in chemical structure on these processes by observing their effects on partitioning in some other system.

These ideas were supported by extensive results obtained by Collander[8] who showed that the partition coefficients for compounds in one pair of solvents can often be satisfactorily related to partition coefficients for another pair of solvents. With organic acids, for example,

$$\log P_{isopent} = 1.17 \log P_{isobut} - 0.17$$

$$\log P_{oct} = 1.24 \log P_{isobut} - 0.42$$

where $P_{isopent}$ is the partition coefficient of the compounds between *iso*pentanol and water, and P_{isobut} and P_{oct} are the coefficients between *iso*butanol and water and *n*-octanol and water, respectively.

Similar ideas have been applied in partition chromatography. The movement of a compound in a particular system is usually expressed as the R_F, which is the ratio of the distance travelled by the substance to the distance moved by the solvent front. If A_L is the cross-sectional area of the moving phase and A_S is the cross-sectional area of the stationary phase,

$$R_F = \frac{A_L}{A_L + PA_S},$$

so the partition coefficient,

$$P = \frac{A_L}{A_S}\left(\frac{1}{R_F} - 1\right).$$

The value of $\log[(1/R_F) - 1]$ has been called R_M and the effects of substituents on R_M have been found to be additive.[9] This is to be expected because R_M is $\log P + a$ constant. Because values of $\log P$ for one solvent system (e.g. *iso*pentanol and water) are linearly related to $\log P$ for another system, the change in $\log P$ produced by a substituent in one system should be directly proportional to the change it produces in another. It is convenient to have a standard system for reference and Hansch[10] has used the partition between *n*-octanol and water. Changes in $\log P$ in this system he called π values. For example, the partition coefficient for benzene in this system is 135, for chlorobenzene it is 692 and for nitrobenzene it is 71; the corresponding values of $\log P$ are 2.13, 2.84 and 1.85, respectively, so the value of π for the chloro group is 0.71; and for the nitro group it is -0.28. Positive values of π indicate increased solubility in the octanol phase.

Some values are listed in Table 2.4.[11] In practice it is found that to some extent they are different in different types of molecule. The part of the molecule attached to the aromatic ring is largely

TABLE 2.4: π *Values (octanol-water)*

	o	*m*	*p*
NO_2	-0.23	0.11	0.24
CN	—	$-0.30\,(-0.24)$	$-0.32\,(+0.14)$
F	0.01 (0.25)	0.13 (0.47)	0.15 (0.31)
Cl	0.59 (0.69)	0.76 (1.04)	0.70 (0.93)
Br	0.75 (0.89)	0.94 (1.17)	1.02 (1.13)
I	0.92 (1.19)	1.15 (1.47)	1.26 (1.45)
HO_2C	—	$-0.15\,(+0.04)$	(0.12)
Me	0.68	0.51 (0.56)	0.52 (0.48)
MeO	-0.33	0.12 (0.12)	$-0.04\,(-0.12)$
HO	—	$-0.49\,(-0.66)$	$-0.61\,(-0.87)$
H_2N	—	(-1.29)	(-1.63)

Values in parentheses—phenols; all others—phenoxyacetic acids

(From results of Fujita, Iwasa and Hansch, *Journal American Chem. Soc.* (1964), no. 86, p. 5177.)

responsible for solubility in the aqueous phase and it is clear that the effect of the substituent on this group is different, for example, in phenols from its effect in phenoxyacetic acids. Correlation of partition coefficients between different systems, however, is usually good. Iwasa, Fujita and Hansch,[12] for instance, obtained a good correlation between π values and the effects of the substituents on R_M in partition chromatography. Such problems as do arise nearly all involve interactions between the substances and water, such as the formation of hydrates, or association in the nonaqueous phase, such as the formation of dimers by organic acids.[13]

EXAMPLE 2:14.

From the pK_a values of substituted benzoic acids listed in Table 2.3 calculate the Hammett σ values for the *m*-chloro, *m*-nitro, *m*-methyl, and *p*-amino groups. From the pK_a values of *m*-chloraniline (3.52) and *m*-toluidine (4.70) calculate the pK_a values of *m*-nitraniline and *p*-phenylene diamine and compare them with the values in Table 2.3 (*m*-toluidine = *m*-methylaniline: *p*-phenylenediamine = *p*-aminoaniline).

The pK_a value of phenol is 9.98 and that of *m*-chlorophenol is 9.02; calculate the values for *m*-cresol ($=$*m*-methylphenol), *m*-nitrophenol and *p*-aminophenol. Comment on the values of ρ for the ionisation of anilines and phenols.

Plot the pK_a values of substituted anilines against those of substituted benzoic acids listed in Table 2.3. Do the results justify using the Hammett relationship?

EXAMPLE 2:15.

The pK_a of phenethylamine ($PhCH_2CH_2NH_2$) is 9.74 at 25°C and 9.46 at 37°C. Calculate the changes in enthalpy (ΔH) and entropy (ΔS) associated with ionisation.

EXAMPLE 2:16.

The partition coefficient of phenol between *n*-octanol and water is 29. Calculate the partition coefficients of *m*-chlorophenol, *m*-iodophenol, and *m*-hydroxyphenol (resorcinol) from the values of π listed in Table 2.4. The partition coefficients of *m*-chlorophenoxyacetic acid and *m*-nitrophenoxyacetic acid in the same system are 178 and 29, respectively. Calculate the partition coefficients for phenoxyacetic acid and for the *m*-iodo and *m*-hydroxy compounds.

EXAMPLE 2:17.

Values of π for substituted phenols are shown below, together with values of ΔR_M in the system triethylene glycol/di-*iso*propyl ether. Estimate the values of a and b in the equation $\Delta R_M = a\pi + b$ (draw a graph and fit the results by eye or use a least-squares-fit if possible); what is the biggest difference between the experimental values of ΔR_M and the values represented by the equation?

XC_6H_4OH	ΔR_M	π
$4\text{-}NH_2$	1.762	-1.63
$3\text{-}NH_2$	1.720	-1.29
$4\text{-}OH$	1.450	-0.84
$3\text{-}OH$	1.374	-0.66
$4\text{-}CN$	0.488	0.14
$3\text{-}CN$	0.488	0.24
$4\text{-}CH_3$	-0.165	0.48
$3\text{-}CH_3$	-0.165	0.50
$4\text{-}NO_2$	0.281	0.50
$3\text{-}NO_2$	0.176	0.54
$4\text{-}Cl$	-0.165	0.93
$3\text{-}Cl$	-0.165	1.04

(From results of Iwasa, Fujita and Hansch, *Journal Med. Chem.* (1965), no. 8, p. 150.)

Thermodynamic Implications of the Hammett Relationship

The Hammett relationship can be treated as empirical but it is reasonable to look for some explanation. A linear relationship between values of $\log K$ for different equilibria can be explained by supposing that the change in $\log K$ produced by introducing a substituent is the result of changes in the free energy of reaction which depend both on the nature of the group and on the nature of the reaction. If K_0 is the equilibrium constant for the unsubstituted compound and K that for one with a substituent,

$$-\Delta G = RT \ln K_0 \quad \text{and} \quad -(\Delta G + a) = RT \ln K$$

so

$$-\frac{a}{RT} = \ln K - \ln K_0 = \ln\left(\frac{K}{K_0}\right).$$

Hammett[14] suggested that the increment in free energy,

$$a = \frac{A}{d^2}\left(\frac{B_1}{D} + B_2\right)$$

where d is the distance of the substituent from the reacting group, A, B_1 and B_2 are constants independent of temperature and D is the dielectric constant of the solvent. The constant A depends on the nature and position of the substituent whereas B_1 and B_2 depend on the nature of the reaction. If the values of K_0 and K refer to a reaction, such as the ionisation of anilines, which is being correlated with the standard reaction, the ionisation of benzoic acids,

$$\log\left(\frac{K}{K_0}\right) = \sigma\rho = \frac{-a}{2.3RT}$$

and if $\sigma = -A/2.3R$

$$\rho = \frac{1}{d^2 T}\left(\frac{B_1}{D} + B_2\right).$$

This breaks up the increment in free energy into contributions from the group and from the nature of the reaction but it is difficult to understand why the relations hold as well as they do because the free energy itself depends on two unrelated properties, the enthalpy and the entropy.

For the Hammett relation to work satisfactorily, changes in chemical structure would have to alter the enthalpy and entropy of the reaction in such a way that the difference, $\Delta H - T\Delta S$, produces the necessary regular increments in free energy. This could occur if both the entropy and enthalpy are altered in a coordinated way, which does not seem likely, or if either the entropy or the enthalpy is relatively constant. There are physical processes in which it seems probable that changes in chemical structure do not alter the entropy of the reaction to any great extent but difficulties may be expected with systems involving water, particularly if the changes in chemical structure involve groups which interact with water in different ways. It would seem, therefore, unlikely that the Hammett relation could be extended to reactions involving highly ordered biological material but attempts have been made to do this and these are discussed in Chapter 5.

Information is needed about entropy or enthalpy changes in reactions, as well as about free energy changes. Sometimes the enthalpy can be measured directly by calorimetry but usually the amount of heat is too small for this to be possible with reasonable amounts of material and it is necessary to try to calculate enthalpy changes by observing the effects of temperature on the reaction. Though this may be possible for chemical processes, it may be very difficult for biological ones. Not only is there often a narrow range of temperature in which the biological process can occur but there is also the likelihood that any change at all may affect numbers of processes involved in a complex biological system in different ways.

References

1. T. Teorell, *Arch. int. Pharmacodyn.* (1937), no. 57, p. 205; ibid., p. 226. A. Goldstein, L. Aronow and S. M. Kalman, *Principles of Drug Action*, 2nd edn, Chapter 4, especially pp. 333–6 (Wiley, New York, 1974).

2. L. S. Goodman and A. Gilman, *The Pharmacological Basis of Therapeutics*, 3rd edn p. 117 (Macmillan, New York, 1965): the partition coefficient between chloroform and buffer at pH 7.4 is 102, S. Mayer, R. P. Maickel and B. B. Brodie, *Journal Pharmacol.* (1959), no. 127, p. 205.

3. M. Davson, *A Textbook of General Physiology*, 2nd edn, p. 226 (Churchill, London, 1959).

4. A. Albert and E. P. Serjeant, *The Determination of Ionization Constants*, pp. 28, 29, 80 (Chapman and Hall, London, 1971).

5. J. Armstrong and R. B. Barlow, *British Journal Pharmacol.* (1976), no. 57, p. 501.

6. L. P. Hammett, *Chem. Review* (1935), no. 17, p. 125.

7. G. Kortüm, W. Vogel and K. Andrussov, *Dissociation Constants of Organic Acids in Aqueous Solution* (Butterworth, London, 1961); D. D. Perrin, *Dissociation Constants of Organic Bases in Aqueous Solution* and Supplement (Butterworth, London, 1965 and 1972).

8. R. Collander, *Acta Chem. Scan.* (1951), no. 5, p. 774.

9. E. C. Bate-Smith and R. G. Westall, *Biochim. Biophys. Acta* (1950), no. 4, p. 427.

10. C. Hansch, P. P. Maloney, T. Fujita and R. M. Muir, *Nature* (1962), no. 194, p. 178.

11. T. Fujita, J. Iwasa and C. Hansch, *Journal American Chem. Soc.* (1964), no. 86, p. 5175.

12. J. Iwasa, T. Fujita and C. Hansch, *Journal Med. Chem.* (1965), no. 8, p. 150.

13. A. Leo, C. Hansch and D. Elkins, *Chem. Review* (1971), no. 71, p. 525.

14. L. P. Hammett, *Journal American Chem. Soc.* (1937), no. 59, p. 96.

CHAPTER 3

The Size and Shape of Molecules

Size: Molal Volumes and Apparent Molal Volumes

Some indication of the size of molecules is given by their molecular weight but more precise information can be obtained from measurements of density. For a pure liquid

$$\text{density} = \frac{\text{the weight of 1 mole}}{\text{the volume of 1 mole}} = \frac{M}{V_m}$$

where M is the molecular weight and V_m is the *molal volume*. If the density can be measured to 4 decimal places, the molal volume can be calculated to $0.1\,cm^3$, and by comparing values of V_m for different compounds, at the same temperature, it is possible to obtain an idea of the size of various groups.[1] Some values are listed in Table 3.1.

TABLE 3.1: *Additive Increments* $(\Delta V_m, cm^3\,mol^{-1})$ *for Molal Volumes at 20°C.*

CH_2	16.6	CHO	25.1
CH_3	31.5	CO (ketone)	10.2
$H_2C{=}CH—$	41.5	COOH	27.2
$HC{\equiv}C—$	33.2	COO— (ester)	19.5
Ph	74.6	SH	10.8
α-Naphthyl	107.3	—S—S—	27.5
β-Naphthyl	109.5	NH_2	17.7
cyclopentyl	79.2	CN	22.7
cyclohexyl	93.9	NO_2	24.5
F	15.1	2-Pyridyl	67.7
Cl	23.0	3-Pyridyl	66.5
Br	26.2	4-Pyridyl	66.3
I	32.9	2-Thienyl	65.6
OH	10.2		
—O— (ether)	6.7		

(Values selected from Exner, *Coll. Czech. Chem. Comun.* (1967), no. 32, p. 1.)

Most drugs are usually given in solution in water, not as pure liquids, and the apparent size in solution can be calculated from measurements of the densities of solutions of known composition by weight (molality). The density can be calculated simply by weighing a known volume of solution (in a glass vessel called a pycnometer) but equipment is now available for measuring density much more conveniently from physical properties dependent on mass.

The *apparent molal volume*, ϕ_v, can be calculated by subtracting the volume occupied by the solvent from the total volume of solution containing 1 mole of solute. If the solute is made up in a concentration of m molal (m mol/kg solvent), there are $1/m$ kg solvent for each mole of solute. The density of the solution,

$$d = \frac{\text{wt solute} + \text{wt solvent}}{\text{total volume}}.$$

If the weights are in grams, the volume should be in cm^3. The total volume is

$$\frac{M + 1000/m}{d}$$

where the molecular weight, M, is in grams. The volume occupied by solvent will be $1000/md_s$ where d_s is the density of the solvent, so the apparent molal volume,

$$\phi_v = \frac{1}{m}\left(\frac{Mm + 1000}{d} - \frac{1000}{d_s}\right).$$

The apparent molal volume varies with temperature and with concentration. Accurate temperature control is needed and if a range of concentrations has been studied it is possible to extrapolate to infinite dilution. For ionised compounds, however, the relation between ϕ_v and molar concentration c includes a term involving $c^{1/2}$,

$$\phi_v = \phi_v^0 + Sc^{1/2} + jc$$

and for uni-univalent electrolytes in water at 25°C the value of S calculated from Debye-Hückel theory is 1.868; j is a constant determined by the solute.[2] The value at infinite dilution, ϕ_v^0, is therefore an estimate of the size of the solute in solution, assuming

TABLE 3.2: *Increments in Apparent Molal Volume at Infinite Dilution ($\Delta\phi_v^0$, $cm^3\ mol^{-1}$) for Solutions in Water at 25°C. (Some increments in V_m from Table 3.1 are included for comparison.)*

	$\Delta\phi_v^0$	ΔV_m
Additional CH_2	16.2	16.6
Replacement of H by Ph	63.4	74.6
Replacement of Ph by cyclohexyl	16.4	19.3
Replacement of H by cyclohexyl	79.1	93.9
Replacement of CH_2 by O	−10.6	−9.9
Replacement of CH_2 by CO	−3.6	−6.4
Replacement of CH_2CH_2 by COO	−14.0	−13.6
Replacement of H by OH	2	
	(very variable)*	
Replacement of Ph by indole	22.1	
Replacement of Cl^- by Br^-	6.9	
Replacement of Br^- by I^-	11.5	
Replacement of $-\overset{+}{N}Me_3$ by:		
Pyridinium	2.2	
$-\overset{+}{N}Me_2Et$	15.0	
Methylpyrrolidinium	19.2	
$-\overset{+}{N}MeEt_2$	30.0	
Methylpiperidinium	33.4	
Ethylpyrrolidinium	34.4	
Quinuclidinium	36.6	
$-\overset{+}{N}Et_3$	44.5	
Ethylpiperidinium	48.0	

* The effects range from −1.1 to +4.3.
(Values from Barlow, Lowe, Pearson, Rendall and Thompson, *Mol. Pharmacol.* (1971), no. 7, p. 357; Millero, *Chem. Review* (1971), no. 71, p. 147; Abramson, Barlow, Franks and Pearson, *British Journal Pharmacol.* (1974), no. 51, p. 81; Barlow, *British Journal Pharmacol.* (1974), no. 51, p. 413; and some unpublished results.)

that it does not alter the volume of the solvent. For ionised compounds it is difficult to estimate the size of the separate ions (e.g. Cl^-) but from values for series of compounds it is possible to estimate changes in size with considerable accuracy. Some values are shown in Table 3.2 and it is interesting to compare them with the corresponding values of the changes in V_m, so as to obtain some idea of interactions between groups and water.

It is sometimes necessary to know the relation between molarity, c, and molality, m. If the density of the solution is d and the molecular weight is M,

$$c = \frac{1000md}{1000 + mM}.$$

EXAMPLE 3:1.

The table below lists densities at 25°C:

Methanol	0.7866	Methylacetate	0.9282
Ethanol	0.7851	Ethylacetate	0.8941
$Me_3CCH_2CH_2OH$	0.8090	3, 3-Dimethylbutylacetate	0.8655

Calculate the molal volumes (V_m) and the size of the acetate group in the three esters. Compare this with the size calculated from the apparent molal volumes at infinite dilution (ϕ_v^0 in cm^3 mol^{-1}) of choline bromide (130.2) and acetylcholine bromide (163.9).

(From results of Barlow and Tubby, *British Journal Pharmacol.* (1974), no. 51, pp. 95–100.)

Shape

The most accurate information about the shape of a molecule is obtained by analysing the diffraction patterns obtained when it is exposed to radiation whose wavelength is of the same order as the distance between the atoms. X-rays with a wavelength of 1–10 Å (0.1–1.0 nm, see page 34) are suitable, though they do not interact appreciably with hydrogen atoms. Mathematical analysis of the diffraction pattern makes it possible to reconstruct the arrangement in space of the atoms giving rise to it. To do this it is usually necessary to have the substance in the crystalline state for the system to be ordered enough, but if the substance is itself a highly ordered macromolecule, such as an enzyme protein, it may be possible to work with strands of wet material.

In simple molecules the standard error of estimates of the distance between atoms in the molecule may be as low as ± 0.002 Å but in larger molecules it is usually about ± 0.02 Å, with a bigger uncertainty about distances involving hydrogen atoms. Some average interatomic distances and bond angles are listed in Table 3.3.[3] Such average distances can be used to construct scale models which will give some idea of the shapes of new compounds, provided these do not contain chemical features which make any of their bonds markedly different from average.

Some types of molecular model attempt to give an indication of the space occupied by atoms, which is obviously important because it may limit the arrangements of the groups which are possible. It is difficult to know exactly what this space is and it can only be

TABLE 3.3: *Covalent Radii (Å) and Bond Angles.*

Covalent radii	Single bond	Double bond	Triple bond
C	0.772	0.667	0.603
N	0.70	0.63 (azo)	0.55
P	1.10	1.00	
O	0.66	0.55	
S	1.04	0.94	
F	0.64		
Cl	0.99		
Br	1.14		
I	1.33		
C in aromatic systems	0.69		
N in C—N bond of amides	0.56		

Bond angles

Tetrahedral angles in 4-covalent carbon, quaternary nitrogen and pentacovalent phosphorus

$$\diagdown\!\!-\!\!\underset{\diagup}{P} = 0, \quad 109\tfrac{1}{2}°.$$

Angles in carbonyl (aldehydes, ketones, esters), $115°$, $122\tfrac{1}{2}°$, $122\tfrac{1}{2}°$; in amides, $114°$, $121°$, $125°$.

Angles between bonds in divalent oxygen, $110°$; in divalent sulphur, $104°$.

*$1\text{Å} = 100\,\text{pm} = 0.1\,\text{nm} = 10^{-10}\,\text{m}$.

(Values from Sutton, *Interatomic Distances* (Chemical Society, London, 1958); Supplement (1965); Dreiding, *Helv. Chim. Acta* (1959), no. 42, p. 1339.)

guessed from the way atoms behave in situations where space is important. Most models use estimates of atomic size based on the *van der Waals' radius*. In the modification of the equation for a perfect gas, van der Waals proposed corrections to allow for the interaction between molecules (van der Waals' forces) and for their volume, and the relation becomes:

$$\left(p + \frac{a}{v^2}\right)(v - b) = nRT$$

where p, v, n, R and T are the pressure, volume, number of moles, gas constant and temperature respectively, and a and b are constants. Some idea of the size of the molecule can be obtained from the constant b, but various assumptions have to be made if this is converted into a 'radius' for use in space-filling molecular models. Another way of assessing this is by observing the closest distance between two atoms in the crystal when there is no chemical bond between them. For example the closest distance between non-bonded bromine atoms in solid bromine is 3.9 Å, so half this value may be taken as the van der Waals' radius of bromine. Although this may give a rough idea of the sizes of atoms or groups, these cannot be known with anything like the same degree of precision that is attached to estimates of the distances between atoms in a molecule and the angles between bonds. It is therefore desirable to obtain additional information from increments in molal volume or apparent molal volume.

Isomerism

Isomers have the same empirical formula and molecular weight but their atoms are arranged differently. If they are arranged in quite different groups, e.g. chloral, Cl_3CCHO, and dichloroacetylchloride, $Cl_2HCCOCl$, they are completely different compounds but if they contain the same groups in different positions, e.g. 1-chloropentane and 2-chloropentane, the chemical differences are much smaller. If the differences in arrangement involve asymmetry, the compounds are described as *optical isomers* because they have different effects on plane-polarised light. Usually it is a carbon atom to which four different groups are attached which gives rise to the asymmetry and makes the compounds optically active. There are two possible arrangements which are mirror-images of each other and are quite distinct; one cannot be superimposed on the other however the molecules are turned around (Figure 3.1). The two forms are called *enantiomers* and the word *chiral* is also used to describe an arrangement such as this which lacks symmetry.

If there are n asymmetric carbon atoms in a molecule there are theoretically 2^n possible optical isomers but not all these are mirror-image forms. A compound with two asymmetric carbon

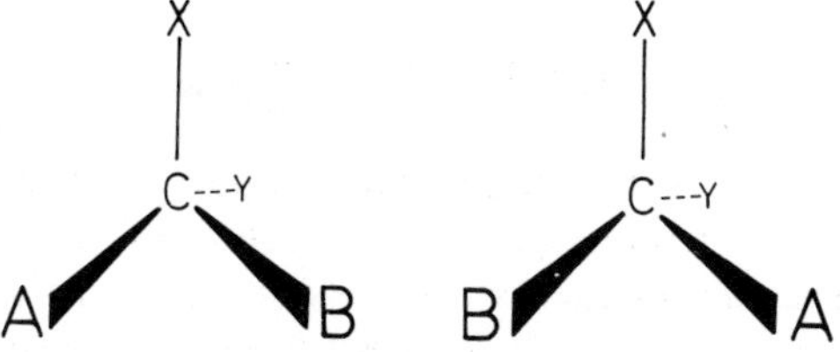

FIGURE 3.1 Two mirror image arrangements (enantiomers) of the compound CABXY. To convert one into the other it is necessary to interchange two groups, e.g. A with B, or B with Y, or X with Y, etc. If two pairs of groups are interchanged, e.g. A with B and X with Y, the original configuration is produced. The thickened lines indicate that the groups are in front, dotted lines that they are behind.

atoms, for example, can have two isomers which have the same arrangement at one atom but a mirror-image arrangement at the other. These are referred to as diastereoisomers; the enantiomers have mirror image arrangements at both atoms. In some situations the introduction of an extra asymmetric carbon atom may lead to forms which contain symmetry. The molecule

$$A \quad\quad\quad A$$
$$\diagdown \quad\quad \diagup$$
$$B-C-C-B$$
$$\diagup \quad\quad \diagdown$$
$$Y \quad\quad\quad Y$$

for instance, exists only in two, not four, optically active forms. The two halves are equivalent so two of the forms are identical and contain a plane of symmetry, which destroys their optical activity (Figure 3.2). If the group Y is replaced by Z at one carbon only, there are four optical isomers.

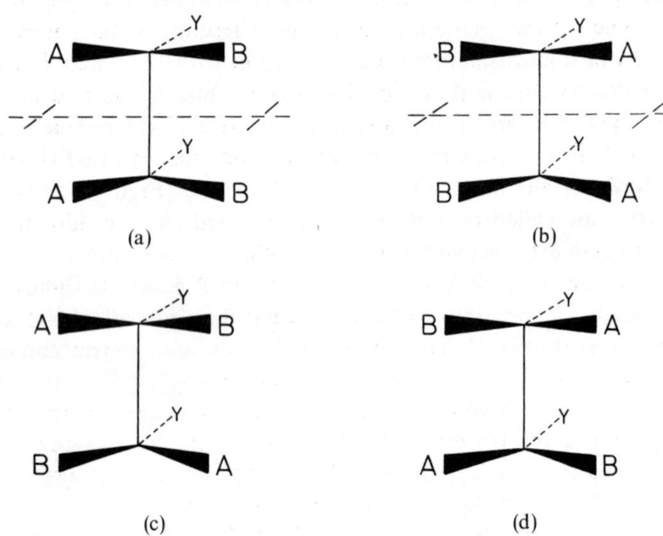

FIGURE 3.2 The four arrangements of

$$A \quad\quad\quad A$$
$$\diagdown \quad\quad \diagup$$
$$B-C-C-B;$$
$$\diagup \quad\quad \diagdown$$
$$Y \quad\quad\quad Y$$

(a) and (b) are identical because they contain a plane of symmetry.

Optical Rotation

The effect on plane-polarised light can be observed when a solution of the compound, or the compound itself if it is a liquid, is placed in a polarimeter. This contains a device, usually a Nicol prism but sometimes a piece of 'polaroid', which filters out light except for rays polarised only in one particular plane. A second similar device is placed at the other end of the light path and rotated so that the plane of the vibrations it will accept is at $90°$ to the plane of vibration of the incident light, so no light should pass through. The sample is placed in the light path between the two devices. If it is optically active it will rotate the plane of polarisation of the light and the second device must be rotated if no light is to pass through. If the sample rotates the plane of polarisation clockwise it is called dextrorotatory or $(+)$; if the rotation is anticlockwise it is called laevorotatory or $(-)$. The extent of the rotation depends on the substance, its concentration, the distance the light passes through the sample, the temperature, the solvent and the wavelength of the light. The *specific rotation*, $\alpha = \theta/lc$ where θ is the observed rotation in degrees, c is the concentration in $g\,ml^{-1}$ and l is the length of the light path in decimetres. The temperature and wavelength are indicated thus: $[\alpha]_D^{20}$ or $[\alpha]_{589}^{20}$ (D indicates the sodium D-line, 589 nm); the concentration (usually in g/100 ml) and the solvent are often included thus: $[\alpha]_D^{20}$ (c, 1.5; water) because the specific rotation is not truly independent of concentration and depends on the solvent.

The *molar rotation*, $[M] = \alpha(M/100)$ where M is the molecular weight, and $[M] = 10\theta/lc$ where c is in mol/l^{-1}. The experimentally observed angle is often very small. With a 2×10^{-2} M solution and a 0.5 dm (5 cm) light path the molar rotation $[M]$ will be the same in degrees as the observed rotation θ measured in millidegrees. For compounds with only one asymmetric carbon atom and light in the visible region $[M]$ may be $20°$ or less, so the angle to be measured, θ, may be less than $0.02°$ in these conditions.

Note that it has been proposed that the symbols for specific rotation and molar rotation should be replaced by α_m and α_n, respectively. The size of the rotation becomes much greater when the wavelength of the light is nearer to that absorbed by groups attached to the asymmetric atom. For most substances, therefore, rotations in the ultraviolet are very much higher (10 times or more) than those in the visible but in some instances the sign of rotation changes. $(+)$-Tartaric acid, for example, is strongly $(-)$ at wavelengths less than 300 nm and has zero rotation at about 350 nm (see Example 3:2). The maximum positive rotation is at about 400 nm. The change of rotation with wavelength is called the *Cotton effect* (after its discoverer) and $(+)$-tartaric acid is described as showing a (weak) positive Cotton effect because the rotation passes through a (weak) maximum as the wavelength is reduced.

One practical complication of working in regions where there is strong absorbance is that little light emerges onto the photocell, which may make it difficult to obtain accurate readings unless a powerful light source is used.

A pair of enantiomers (mirror image forms) should have specific rotations of equal size but opposite sign. Problems arise because it is difficult to obtain one form entirely free from the other. If the rotation, α, of the pure form is known, the observed rotation ϕ can be used to calculate the stereochemical purity, p. There will be a fraction $1-p$ of the other enantiomer present, with a specific rotation of $-\alpha$ so $\phi = p\alpha - (1-p)\alpha = 2p\alpha - \alpha$ and $p = (\phi+\alpha)/2\alpha$; for a racemic mixture $p = 0.5$ so $\phi = 0$.

Stereochemical Purity and Biological Activity

If the biological activity of two enantiomers is additive (pp. 156, 160) the activity of mixtures of optical enantiomers can be treated similarly to the rotation. If the activity of one isomer, when pure, is 1 and that of the other is x (where $x < 1$), the activity of a mixture containing a fraction p of the first (stronger) isomer will be $p+(1-p)x$.

For a racemic mixture $p = 0.5$ so the activity, r, of the racemate will be

$$0.5 + 0.5x = \frac{1+x}{2}.$$

If the weaker isomer is inactive, $x = 0$ and $r = 0.5$; if the two isomers are equiactive, $x = 1$ and $r = 1$. The activity of the racemate should therefore lie between 1 and 0.5 times that of the stronger isomer. It sometimes happens that the stronger isomer has been purified and the racemate compared with it, but attempts to purify the weaker isomer have failed. It should be possible to calculate its activity from that of the racemate because $r = (1+x)/2$ so $x = 2r-1$ and if r is 0.6, $x = 0.2$ and the ratio of the activity of the strong isomer to the weak one, the *stereospecific index*, is $5:1$. If r is 0.55, $x = 0.1$ and the ratio is $10:1$. The value of r cannot be less than 0.5, however, if the activities of the isomers are additive, and it is difficult to make accurate enough estimates of the activity of the racemate to be sure that the value of r is 0.505 (stereospecific index 100) rather than 0.5005 (stereospecific index 1000); see Example 3:3.

If the weaker isomer has been obtained pure but the stronger isomer has only a purity p, its activity will be $p+(1-p)x$, compared with x for the weaker isomer. The stereospecific index will therefore be $(p/x)+1-p$. If the weak isomer is not very strong, x will be small and the expression becomes p/x because $(1-p) \ll p/x$.

Problems arise when it is the weaker isomer which is contaminated with the stronger one; this is also a situation which is commonly found because more of the weaker material will be needed for testing so purification may be less than with the stronger isomer. If p now refers to the purity of the weaker isomer, with an activity x, the apparent activity will be $px+1-p$ and the apparent stereospecific index, R_{obs} will be

$$\frac{1}{px+1-p} \quad \text{so} \quad x = \frac{1-(1-p)R_{obs}}{pR_{obs}}$$

and the true stereospecific index,

$$R = \frac{1}{x} = \frac{pR_{obs}}{1-(1-p)R_{obs}}.$$

For example, if the weaker isomer is 90% stereochemically pure and the apparent stereospecific index is 8, the activity of the weaker isomer is not 0.125 but

$$\left(\frac{1-0.1 \times 8}{0.9 \times 8}\right) = 0.028$$

and the true stereospecific index is 36. Note that the activity must be at least that of the proportion of stronger isomer present. If 10% of this is present, the activity must be at least 0.1, even if the weaker isomer is totally inactive, so the apparent stereospecific index cannot be greater than 10.

If the weaker isomer has been obtained pure and the stronger has not, the true stereospecific index can be calculated from the ratio, RR, of the activity of the racemate relative to the weaker isomer. The activity, r, of the racemate $= (1+x)/2$ so the ratio

$$RR = \frac{r}{x} = \frac{1+x}{2x} \quad \text{and} \quad x = \frac{1}{2RR-1}$$

so the true stereospecific index,

$$R = \frac{1}{x} = 2RR-1.$$

It is very difficult to be sure when the process of purifying stereochemical isomers is complete. In theory it is continued until further purification produces no further increase in the size of the rotation so when two enantiomers have been resolved (stereochemically purified) they should have rotations of the same size but of opposite sign. This could mean that they are both equally stereochemically impure. In this situation the observed stereospecific index,

$$R_{obs} = \frac{p+(1-p)x}{px+1-p}$$

where p is the degree of resolution which has been achieved. Both $(1-p)$ and x should be small (if the isomers differ appreciably in activity) so the expression can be written

$$R_{obs} = \frac{p}{px+1-p}$$

and as the true stereospecific index,

$$R = \frac{1}{x}, \quad \frac{1}{R_{obs}} = \frac{1}{R} + \frac{1-p}{p} \quad \text{or} \quad \frac{1}{R} = \frac{1}{R_{obs}} + \frac{p-1}{p}.$$

If the weaker isomer were completely inactive, the true stereospecific index would be infinite and the expression would become

$$\frac{1}{R_{obs}} = \frac{1-p}{p}, \quad \text{or} \quad p = \frac{R_{obs}}{1+R_{obs}}$$

so the observed stereospecific index can be used to set a limit to the degree of stereochemical purity.[4] When two enantiometers differ appreciably in activity it is remarkable how sensitive an indication of stereochemical purity the stereospecific index can give. For example, an observed index of 100 indicates that the stereochemical purity must be at least $\frac{100}{101} = 99.01\%$.

Although some mirror-image forms differ very greatly in biological activity it is not correct to regard stereospecificity as absolute, with one isomer fitting the receptors and the other not fitting them at all. High stereospecificity is shown, for example, by (+)- and (−)-hyoscyamine (the enantiomeric forms of atropine) at 'muscarinic' acetylcholine receptors. The stereospecific index is over 330 but the weaker (+)-isomer still has affinity comparable with that of the compound lacking the hydroxyl group, the ester of

$$PhCHCOOH$$
$$|$$
$$CH_3$$

instead of

$$PhCHCOOH$$
$$|$$
$$CH_2OH$$

It has been suggested[5] that 'stereoselective' might be a better word to use but the choice should not matter provided the phrase includes 'degree of' or 'index'.

Pairs of enantiomers are often used to test whether uptake processes in tissues are specific or non-specific. If the uptake depends only on physicochemical properties the stereospecific index should be 1 but if some specific carrier mechanism is involved it should be appreciably bigger than this. The value observed, however, will depend on the stereochemical purity of the isomers, as well as on the process itself. This may be difficult to measure if studies are made with small amounts of labelled enantiomers but activity in some other biological test may be used to set limits, even if it is not possible to measure optical rotations.

EXAMPLE 3:2. The rotations, measured in millidegrees, observed with 0.1 M aqueous solutions of (+)- and (−)-tartaric acids ($C_4H_6O_6$) in a cell with a 5 cm light path were:

Wavelength nm	Angle	s.e.	Angle	s.e.
589	+110	2.9	−122	6.2
400	+118	3.3	−127	7.2
360	+34	7.5	−37	2.3
340	−76	9.2	+79	3.8
300	−739	34	+749	8.5
260	−4572	167	+4717	66

Values are the mean and standard error of four estimates and have been corrected for rotation by the cell (slight strain in the silica windows). Do the results indicate a difference in the stereochemical purity of the two samples? Calculate the molar rotations and the stereochemical purity of the less pure isomer, assuming that the other isomer is completely pure. What are the specific rotations at 589 nm?

(Results of a class experiment.)

EXAMPLE 3:3.

A sample of $(\pm)$-nicotine was compared with $(-)$-nicotine for its ability to cause contracture of the chick biventer-cervicis muscle and the equipotent molar ratio was estimated to be 1.77:1. Calculate the activity of the $(+)$-isomer and the stereospecific index.

Subsequently a sample of $(+)$-nicotine was compared directly with $(-)$-nicotine on this preparation and the stereospecific index was estimated to be 4.6. What is the error in the first assay (of the racemate) if this second estimate is correct?

(From results of Barlow (1965), *Tobacco Alkaloids and Related Compounds*, edited by Von Euler, Pergamon, Oxford, p. 288, and Barlow and Hamilton (1965), *Br. J. Pharmacol.*, no. 25, p. 206.)

EXAMPLE 3:4.

Resolved samples of 3-acetoxyquinuclidine had $\alpha_D^{20} - 11.5$ and $+7.7$ (0.1 M in water) and $M_{300}^{20} - 143$ and $+118$. The $(+)$-isomer was 12.7 times as active as the $(-)$-isomer when tested for acetylcholine-like activity on the guinea-pig ileum. If the $(-)$-isomer has been fully resolved, what is the stereochemical purity of the $(+)$-isomer and the ratio of the activities of the pure isomers?

If the weaker $(-)$-isomer had been the one which was incompletely resolved, what would the ratio of the activities of the pure isomers be, supposing that the observed ratio (with incompletely resolved $(-)$-isomer and completely resolved $(+)$-isomer) was (a) 5.7, (b) 6.7 and (c) 11.0.

(From results of Barlow and Casy, *Mol. Pharmacol.* (1975), no. 11, pp. 690–3.)

EXAMPLE 3:5.

A sample of (−)-hyoscine methiodide had a molar rotation of −1184° (300 nm, 20°C) and the estimate of $\log K$ for the muscarinic acetylcholine receptors of the guinea-pig ileum was 9.702. A sample of (+)-hyoscine methiodide had a molar rotation of 1105° (300 nm, 20°C) and the estimate of $\log K$ was 8.622. Calculate the stereochemical purity of the (+)-form, assuming the (−)-form is fully resolved. Estimate the value of $\log K$ for the pure (+)-form; values of K for the enantiomers are additive (see page 160). The stereochemical purity of the (+)-form may have been overestimated because the (−)-form may not have been fully resolved, what is the lowest limit which the stereochemical purity must exceed?

(From results of Barlow, Franks and Pearson, *Journal Med. Chem.* (1973), no. 16, p. 443.)

Description of the Arrangements of Groups about an Asymmetric Carbon Atom

The actual arrangement of the groups around an asymmetric centre is called its *absolute configuration*. Until 1951 no absolute configuration was known and it was only possible to describe the arrangement in terms of the arrangement in other molecules and hence to assign a *relative configuration*. This was achieved by converting one compound into another by reactions which did not involve breaking the bonds attached directly to the asymmetric atom. If these bonds are broken in a reaction there is the distinct possibility that the new group may be attached in such a way that the mirror image arrangement is produced, a process known as the

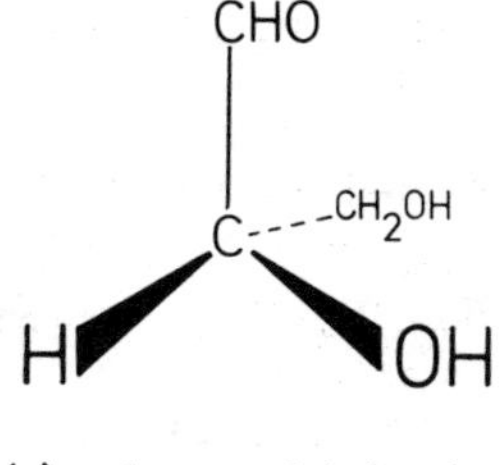

FIGURE 3.3 Structure of D-(+)-glyceraldehyde.

Walden inversion. In some other reactions the product is racemic because it proceeds through a planar intermediate.

All configurations were originally related to (+)-glyceraldehyde and the structure shown in Figure 3.3 was called a D arrangement. The mirror image form was called L.[6] D-(+)-glyceraldehyde can easily be oxidised to glyceric acid without any risk of an alteration in configuration, but this acid is actually laevorotatory. This in turn can be related to lactic acid and so on (Figure 3.4).

In amino acids the OH of compounds related to glyceraldehyde is replaced by NH_2. Although this change could be made chemically, it is likely to lead to an inversion of configuration and so cannot be used to establish configurations. The compound (+)-serine, which is taken as a standard to which the configuration of other amino acids is referred, is assumed to have the same configuration as (+)-glyceric acid, i.e. to be an L-structure (Figure 3.4). Justification for this is obtained by studying the rotations of many derivatives of (+)-lactic acid and (+)-alanine. When esters were made, for example, their rotations were different but the changes in rotation produced by esterification were similar in the two series of derivatives.[7] (+)-Alanine and (+)-serine can be related to each other chemically because the —CH_2OH group can

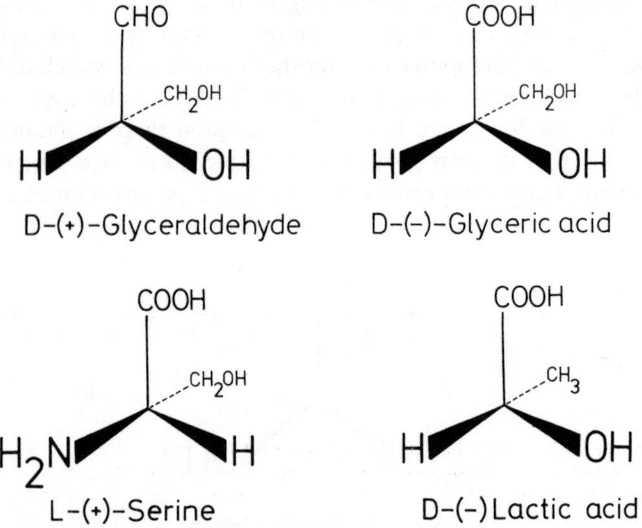

FIGURE 3.4 Relation between (+)-glyceraldehyde, (−)-glyceric acid, (+)-serine and (−)-lactic acid.

be converted into —CH_3 without affecting the bonds attached to the asymmetric carbon atom. Many amino acids can similarly be related to each other chemically but this is not always possible. With phenylalanine, for instance, it is necessary to infer the relationship to alanine by comparing the rotations of series of derivatives.

Instead of working with series of compounds, however, it may be possible to establish relationships by comparing rotations which have been observed over a range of wavelengths, i.e. to compare

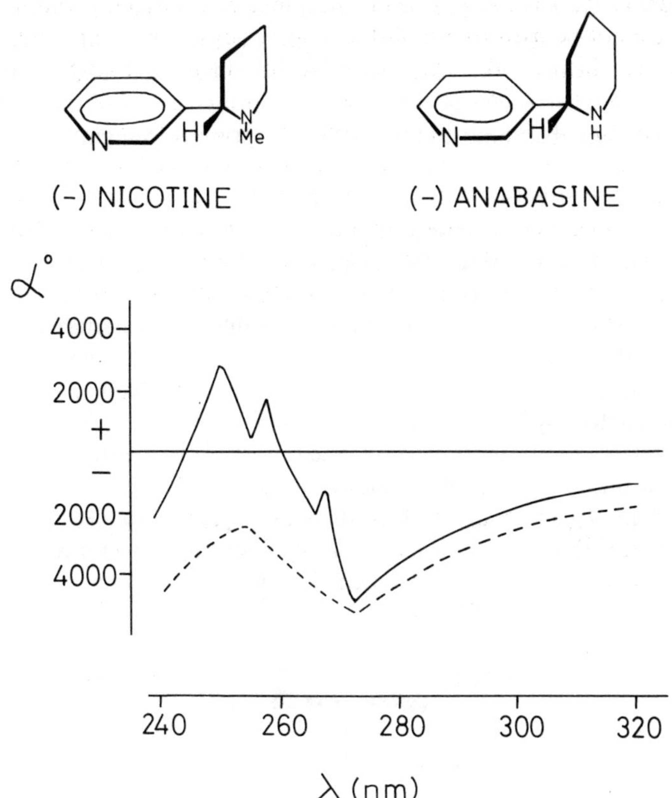

FIGURE 3.5 Optical rotatory dispersion curves for (−)-nicotine (———) and (−)-anabasine (————). These establish that the configuration about the asymmetric carbon atom is the same in both. The (+)-enantiomers have mirror-image curves, i.e. the − and + signs on the *y*-axis are interchanged. The absolute configuration is correctly shown in the formulae.

(Graphs re-drawn from results of Craig and Roy (1965) *Tetrahedron*, no. 21, p. 401.)

optical rotatory dispersion (o.r.d.) curves. The alkaloids anabasine and nicotine, for example, are closely related but cannot be interconverted chemically (Figure 3.5). The o.r.d. curve for (−)-anabasine, however, is very similar to that for (−)-nicotine so it can be concluded that the two have the same arrangement at the asymmetric carbon atom.[8] The mirror-image forms will have mirror-image o.r.d. curves. This method has been used extensively for establishing relative configurations of closely related groups of alkaloids[9] (e.g. (+)-tubocurarine). It is also useful for expressing the coiling of proteins (page 99). Some instruments operate using *circular dichroism* (c.d.) and record differences in the absorption of right and left circularly polarised light instead of rotation. These produce graphs which are effectively the differential coefficient of the o.r.d. curve; effects are only seen in the regions where the substance is absorbing and the optical rotation is changing.

Absolute Configuration—Tartaric Acid

In 1951, Bijvoet, Peerdeman and Van Bommel[10] succeeded in establishing the absolute configuration of (+)-tartaric acid. Normally X-ray diffraction does not make it possible to assign an absolute configuration because the same diffraction pattern would be produced by either of two mirror image arrangements. By studying a structure containing a heavy atom, rubidium, which is itself excited by X-radiation of a suitable wavelength, however, a phase-lag was produced which made it possible to identify uniquely the positions of the other atoms (except hydrogen) in sodium rubidium (+)-tartrate. It was therefore possible to describe the absolute configuration of the naturally occurring (+)-tartaric acid from which this salt was obtained. This is shown in Figure 3.6 and

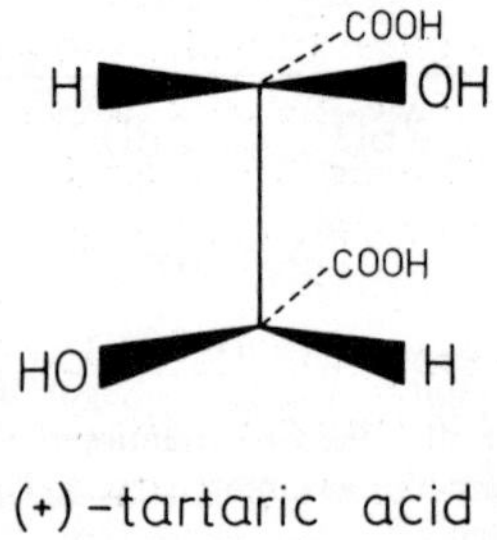

FIGURE 3.6   Absolute configuration of (+)-tartaric acid.

can, unfortunately, be described in two ways. It can be described as being related to glyceraldehyde, if the COOH group is reduced to CH_2OH, and the arrangement would be L at both asymmetric carbon atoms. It can also be regarded as being related to serine if the OH group is replaced by NH_2, but this arrangement would be D. The confusion arises because the chemist looks at the arrangement of different sets of groups, H, OH and CH_2OH in sugars and H, NH_2 and COOH in amino acids. As long as compounds contain these groups there is no confusion but once the relation is extended to compounds containing groups 'like' these groups there may be more than one way of thinking what is like what (see below).

The discovery of the absolute configuration of $(+)$-tartaric acid showed that Fischer was correct in the structure he assigned to it.[6] The arrangement in $(+)$-tartaric acid, supposing the COOH group to be CH_2OH, is the opposite of that in D-glyceraldehyde, so in the latter the arrangement must be as shown in Figure 3.3. It also follows that the arrangement of the H, NH_2 and COOH groups in L-$(+)$-serine must also be correct and all configurations expressed relative to these standards are now absolute configurations.

A Sequence Rule

Because of the confusion about the description of absolute configuration by reference to standards, Cahn, Ingold and Prelog[11] introduced a *sequence rule* which assigned priorities to the bonds based on atomic weight. Each of the four bonds about an asymmetric carbon atom is inspected and the lowest priority (d) is given to that involving the atom of lowest atomic weight. The other bonds are then placed in increasing order of priority, $c < b < a$ and the arrangement is regarded as being the steering-wheel of a car where d is the steering-column and a, b and c are spokes of the wheel. The observer imagines himself sitting at the steering wheel (Figure 3.7) and turning it from a to b to c. If this turns the car to the right, the arrangement is described as R; if it turns it to the left the arrangement is S (R and S are chosen to avoid confusion with D and L).

Application of this sequence rule must result in only one description of the arrangement of the groups about the asymmetric centre. If there is no difference in priority in the atoms attached to the asymmetric centre, the inspection passes to the next atoms and so on. For example, with $(+)$-tartaric acid, the hydrogen (atomic

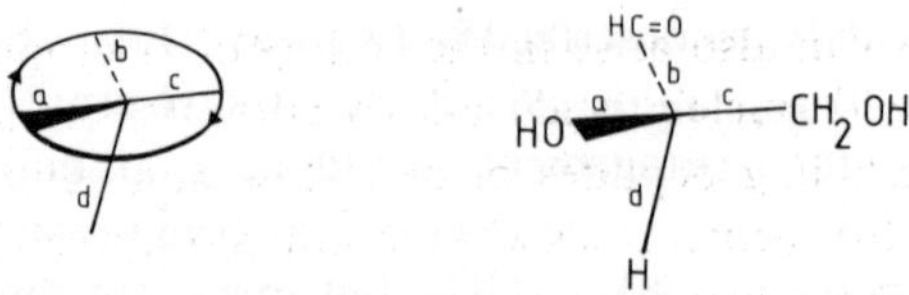

FIGURE 3.7 Assignment of the description R or S. The priorities in (+)-glyceraldehyde are C—OH = *a*, C—H = *d*; the other two bonds are both C—C so the priorities are determined by the groups attached to them. The aldehyde group has two bonds to oxygen and so takes precedence over the CH₂OH, which has only one bond to oxygen. Rotation of the steering wheel from *a* → *b* → *c* will make a right turn, so the assignment is R.

weight = 1) ranks lowest (*d*) and the oxygen (atomic weight = 16) ranks highest (*a*). To decide between the two bonds involving carbon (atomic weight = 12) it is necessary to look at the next atoms to which they are connected. One (COOH) has three bonds to oxygen, whereas the other has only one bond to oxygen so the priority is therefore higher for COOH (*b*) than for CH(OH)COOH (*c*). If the bonds attached to the upper asymmetric carbon atom are considered as a steering-wheel arrangement with the bond to hydrogen as the steering-column, the movement of the wheel from *a* to *b* to *c* will turn the car to the right, so the arrangement is R. The same is true for the arrangement at the lower asymmetric carbon and the full description is RR.

It should be noted that the symbols R and S do not have the 'family' association which D and L have. A change in substituent may well alter the priorities, which are based on atomic weight, so closely related compounds may have opposite symbols. For example, D-(−)-lactic acid has the R configuration but if a chlorine (atomic weight = 35.5) is introduced into the methyl group the assignment of priorities to the carboxyl and methyl groups will become reversed so the chloro compound with the same configuration will be S.

The need for the sequence rule is particularly great in work on drugs which are usually not obviously related to sugars or amino acids. It has become increasingly important as the absolute configurations of more and more compounds are becoming known and in many instances it has now become possible to use the phase-lag from a nitrogen atom to obtain absolute configurations by X-ray diffraction.

Asymmetry can also arise in the coiling of macromolecules such as proteins and nucleic acids and occurs also in much smaller fragments, including peptides of quite low molecular weight. The direction of the helix, clockwise or anticlockwise, is sometimes known from the X-ray diffraction pattern and the symbol P is used to indicate a clockwise (right-handed) arrangement, whereas M is used to indicate an anticlockwise arrangement.[12] The direction of the helix may also be obtained from optical rotatory dispersion but in a molecule in which part is coiled one way and part coiled the other, this method can only give an average indication.[13]

EXAMPLE 3:6.

How many optical isomers are there of the monomethyl ester of tartaric acid?

EXAMPLE 3:7.

The absolute configuration of $(-)$-adrenaline (epinephrine) is known to be:

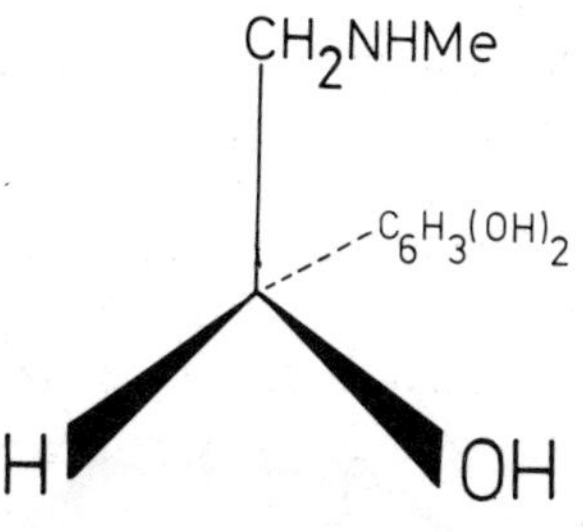

Is this D or L? Is it R or S?

EXAMPLE 3:8.

What is description of the absolute configuration of $(-)$-nicotine and $(-)$-anabasine, shown in Figure 3.5?

EXAMPLE 3:9.

Kainic acid has the structure:

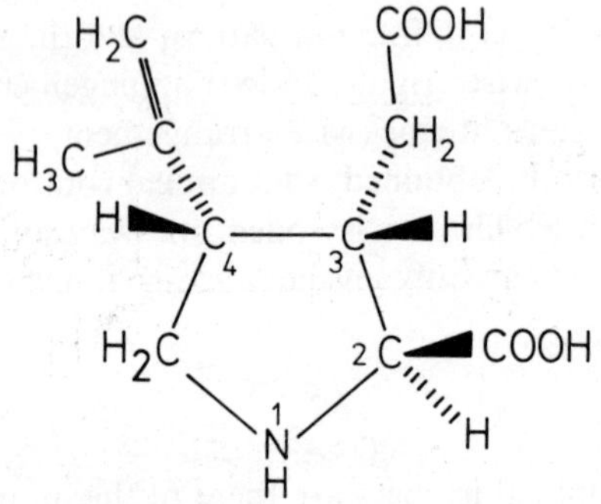

KAINIC ACID

Use the R and S convention to describe its absolute configuration.

EXAMPLE 3:10.

$(-)$-Ecgonine has the structure:

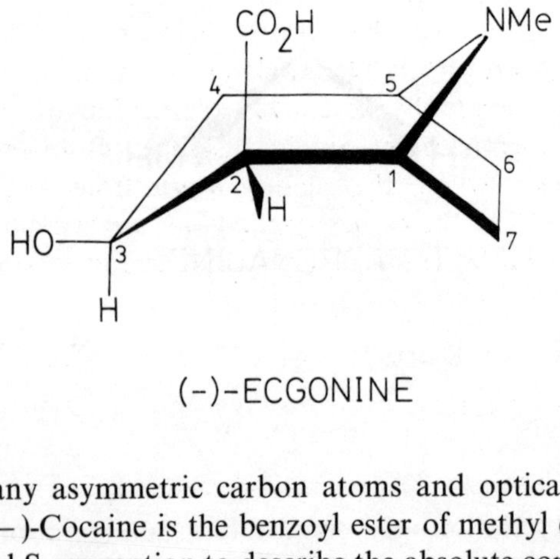

$(-)$-ECGONINE

How many asymmetric carbon atoms and optical isomers are there? $(-)$-Cocaine is the benzoyl ester of methyl ecgonine; use the R and S convention to describe the absolute configuration.

EXAMPLE 3:11.

6-Aminopenicillanic acid has the structure:

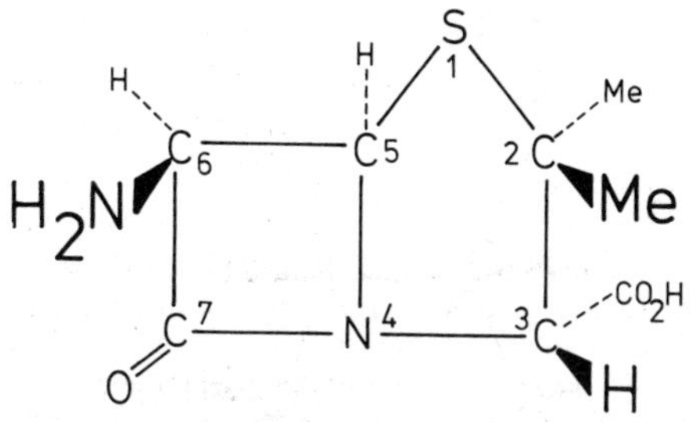
S
1
H
H
Me
C 6
C 5
2 C
Me
H2N
C 7
N 4
3 C
CO2H
O
H

6-AMINO PENICILLANIC ACID

Use the R and S convention to describe the absolute configuration.

Geometrical Isomerism

Rigidity in a molecule may give rise to the existence of two distinct arrangements of groups attached to the part which is stiff. In the simplest examples such as fumaric and maleic acids (Figure 3.8), the groups are attached to a double bond and the physical and chemical properties of the two *geometrical isomers* are markedly different. Maleic acid more readily forms an internal anhydride from which it is obtained back again by adding water so the two carboxyl groups must be close together (*cis*); the *trans* arrangement, however, is thermodynamically more stable and maleic acid can gradually be converted into fumaric acid, which has the higher melting point. Similar instances of isomeric pairs are found in oximes, where the rigidity arises from a C═N bond instead of C═C and the names *syn* and *anti* are sometimes used in place of *cis* and *trans* to describe the arrangements. In these

H-C-COOH
HOOC-C-H
H-C-COOH
H-C-COOH

Fumaric acid Maleic acid

FIGURE 3.8 Examples of geometrical isomers.

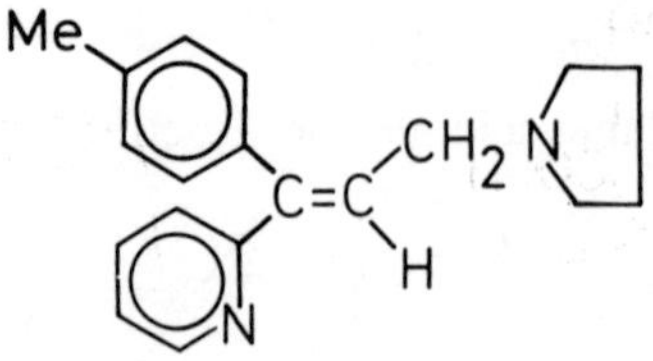

Triprolidine

FIGURE 3.9 The active form of triprolidine. Note that the benzene and pyridine rings cannot lie in the same plane. The form shown has an ultraviolet absorption spectrum similar to 2-vinylpyridine so the pyridine ring and the double bond lie in the same plane with the benzene ring twisted out of the way. Its affinity for histamine receptors is 1170 times that of the isomer with the nitrogen containing groups in the *cis* position (Ison, Franks and Soh (1973), *J. Pharm. Pharmacol.*, no. 25, p. 887).

examples it is fairly easy to see which groups are being referred to by the descriptions (*cis* and *trans*) but this is not always clear. For example the antihistamine triprolidine can exist in two geometrically isomeric forms because it is an allylamine and contains a C=C bond (Figure 3.9). Both isomers are known and one is much more active than the other. The more active can be described as *trans* but this refers to the groups containing nitrogen, i.e. pyrrolidylmethyl and α-pyridyl. To avoid any ambiguity the sequence rule of Cahn, Ingold and Prelog (page 97) may be used to assign priorities and the description applied to the two groups of higher priority. In this instance the pyrrolidylmethyl has higher priority than hydrogen at one end of the double-bond and at the other the nitrogen containing pyridine ring has higher priority than the benzene ring. The symbols Z (*zusammen* = together) and E (*entgegen* = opposite) are then applied to indicate the arrangement[14] which in this instance is E.

Rigidity can also arise because parts of the molecule are held in a ring. Five-membered rings such as cyclopentane, tetrahydrofuran and pyrrolidine are good examples but in them geometrical isomerism is often associated with optical isomerism because the introduction of a substituent in, say, a pyrrolidine ring produces an asymmetric carbon atom and with two groups substituted, the use of the sequence rule to describe the absolute configuration simultaneously describes the geometrical arrangement. For example, the polypeptide antibiotic telomycin contains both *cis* and *trans* 3-hydroxy-L-proline. The carbon atoms in both the 2 and 3

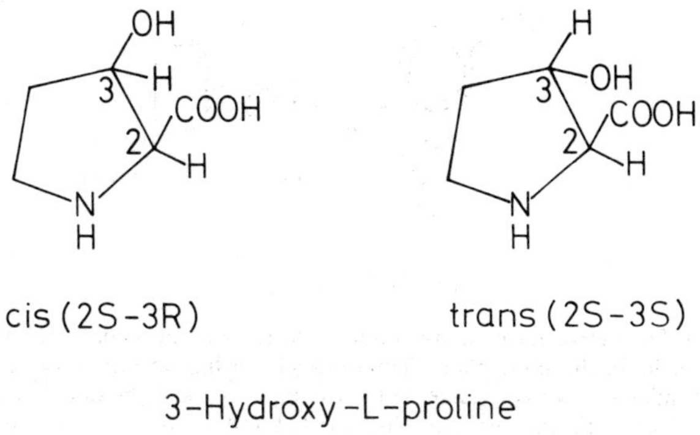

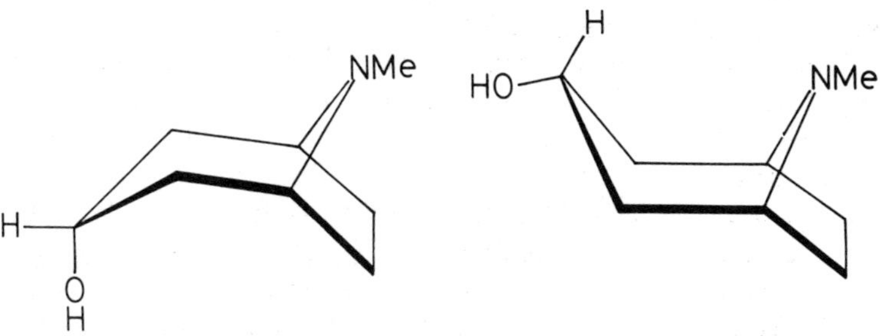

cis (2S–3R)   trans (2S–3S)

3–Hydroxy –L–proline

FIGURE 3.10 Forms of 3-hydroxy-L-proline.

positions are asymmetric, however, and the two arrangements can fully be described by 2S, 3R and 2S, 3S (Figure 3.10); in these two arrangements the COOH and OH groups must be *cis* and *trans* respectively; it is not necessary to say so.

With six-membered rings, such as cyclohexane, tetrahydropyran and piperidine there is the complication that the ring is buckled and that in some positions the molecule produced may have a plane of symmetry so some possibilities for optical isomerism disappear, as in tropine and *pseudo*tropine for instance. In these the piperidine ring has a pyrrolidine ring fused to it which is rigid. The piperidine ring can be either in a *boat* or *chair* form but the latter predominates (Figure 3.11). If a hydroxyl group is introduced into the piperidine ring opposite the nitrogen atom there are two possible arrangements but the molecule possesses a plane of

FIGURE 3.11 'Chair' and 'boat' forms of tropine; in *pseudo*tropine the hydrogen and hydroxyl groups are interchanged (the other hydrogen atoms have been omitted).

symmetry and there is no optical activity. The two forms cannot be described by the sequence rule because two of the bonds have exactly the same priority. In one form (tropine) the hydroxyl group is as far away from the nitrogen atom as is possible, in the other (*pseudo*tropine) it is nearer (and there is the possibility of groups moving from O to N or vice versa when the molecule assumes a 'boat' conformation). The forms might be described as *trans* and *cis* respectively but the term *trans* applied to tropine can be slightly misleading, because it suggests that it might be the more stable. In fact the more stable arrangement is with the hydroxyl group in the same plane as the bulk of the molecule (*equatorial*) rather than projecting away from it (*axial*). The symbol E could be used for tropine and Z for pseudotropine.

When two saturated rings are fused together, the differences between the shapes of geometrical isomers and between their chemical properties become even more marked. With the *cis* and *trans* forms of decalin (decahydronaphthalene; Figure 3.12) there are small but detectable differences in the infrared absorption spectra and this applies also to many of their derivatives, including steroids. The infrared spectra indicate the absorption of energy

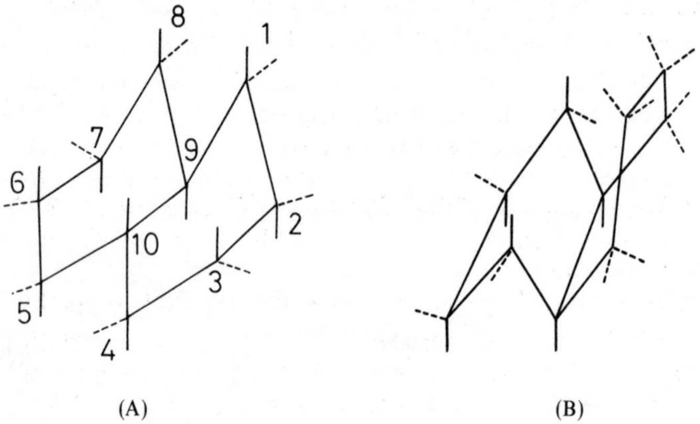

(A)   (B)

FIGURE 3.12 (A) *Trans*-decalin, $C_{10}H_{18}$; 10 'axial' bonds, 8 'equatorial' bonds (shown as broken lines). (B) One form of *cis*-decalin.

The symbol α may be used to describe bonds below the plane of the ring as drawn and β to describe bonds above the plane. These are often used in steroids and prostaglandins but they have no absolute significance; their significance depends on which way up the molecule is drawn. Where two rings are fused, however, as in the decalins, an αβ arrangement (e.g. 9α, 10β) must be *trans* and an αα or ββ arrangement must be *cis*.

associated with the movements of bonds in a molecule[15] (stretching, bending, rocking, etc.). In simpler geometrically isomeric pairs the differences between possible movements may be much smaller and difficult to detect in the spectra but they can clearly be seen in nuclear magnetic resonance spectra where the energy absorbed depends on the atomic environment of the nuclei of the atoms (usually protons or ^{13}C) which are absorbing the radiation (see below). With optical isomers there are no differences in n.m.r. spectra, though some may be produced by the addition of suitable optically active 'shift reagents'.[16]

Because there are differences in the physical properties of geometrical isomers it is easier to obtain them pure than it is to resolve optical isomers. The problem of the presence of one geometrical isomer in a sample of the other should therefore be easier to solve and it should be possible to estimate purity by physical means (especially from n.m.r. spectra). Corrections for contamination can, however, be made in the same way as with optical isomers, provided the biological activities of the two forms are additive. For example, if the activity of the weak isomer relative to the strong one is x (where $x < 1$) and the sample is only a fraction, p, pure, its activity will appear to be $px + (1-p)$ and the ratio R_{obs} will be $1/(px+1-p)$ instead of $1/x$. If p is known, the true value of R $(=1/x)$ can be calculated for $R_{obs}(px+1-p) = 1$ so

$$\frac{1}{R} = \frac{1}{R_{obs}p} + \frac{p-1}{p}.$$

If both isomers are equally impure the relationship is the same as for enantiomers (page 90), but this situation is less likely to occur with geometrical isomers.

Although the difference between the physical properties of geometrical isomers makes them easier to obtain pure than optical enantiomers, it also makes them less suitable as tools for studying pharmacological problems, such as whether storage mechanisms are 'specific' or 'nonspecific'. Differences in biological properties could partly be due to differences in physical properties.

EXAMPLE 3:12.

Prostaglandin E_2 and Prostaglandin $F_{1\alpha}$ have the structure (omitting hydrogen atoms):

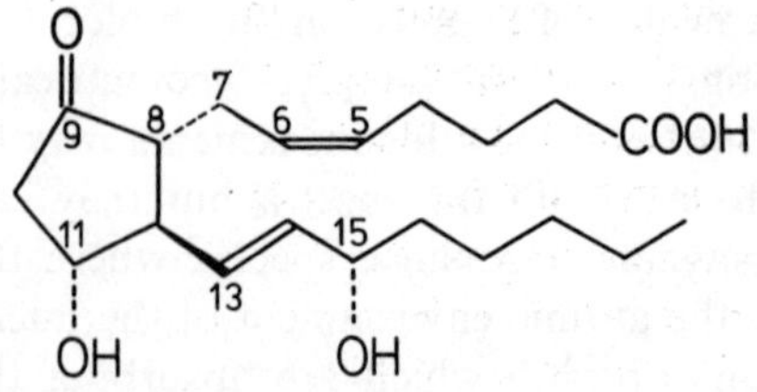

PROSTAGLANDIN E_2

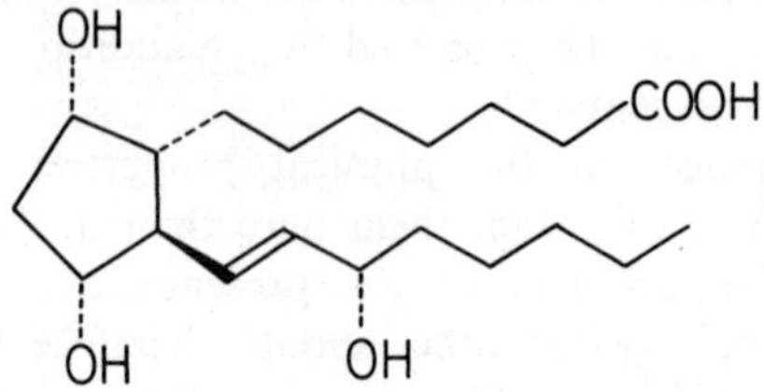

PROSTAGLANDIN $F_{1\alpha}$

How many asymmetric carbon atoms are there in each? Describe their absolute configuration. Will this be altered by the removal of the 5:6 double bond (giving PGE_1) or its insertion into $PGF_{1\alpha}$ (giving $PGF_{2\alpha}$)?

EXAMPLE 3:13.

The structures of the glucocorticoid, cortisone and the mineralocorticoid, aldosterone, are:

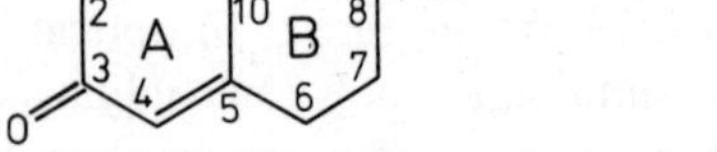

Cortisone

Aldosterone

How many asymmetric carbon atoms are there?

The description α is usually given to groups attached below the plane of the ring as drawn and β to groups above the plane (this applies also to prostaglandins). The B and C rings are fused

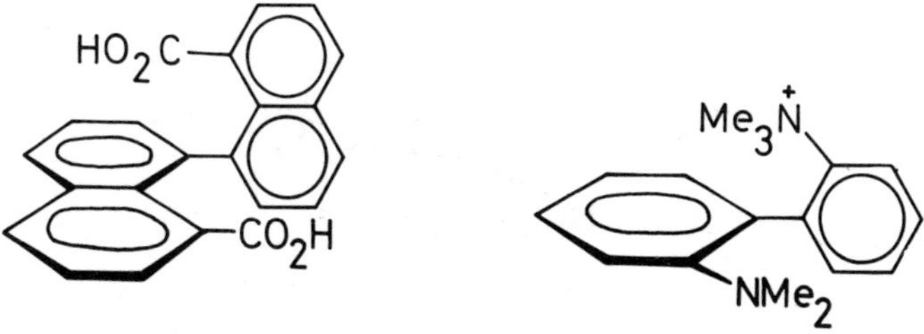

FIGURE 3.13 Sterically hindered biphenyls which have been resolved. These are not flat: the two ring systems cannot be in the same plane.

trans (the hydrogen atoms at the 8 and 9 positions are β and α respectively) and the hydrogen at the 14 position is α in both compounds. Construct models of these compounds, describe their absolute configuration, and compare their shapes.

Preferred Conformations

Although parts of some molecules may be rigid, there should be 'free' rotation about single bonds and many molecules, or parts of molecules, might be expected to be extremely flexible. In fact the freedom to rotate may be relatively limited, particularly if the atoms forming the bond have large groups attached to them which get in each other's way (as in triprolidine, p. 102). In extreme instances, as with some di-*o*-substituted biphenyls, the rotation is so hindered sterically that there are two distinct optical enantiomers (Figure 3.13). Even with simple aliphatic compounds the possible arrangements differ in stability and the molecules are likely to be found in a limited number of conformers.

These arrangements can be represented by *Newman projection formulae*. The bond is viewed end-on and the bond between the nearer atom and the rest of the molecule is set at 12 o'clock (Figure 3.14). In the most stable arrangements, large groups are usually arranged as far away from each other as possible, i.e. *trans* to each other. The *eclipsed* position, with the substituents as close to each other as possible, is much less stable; for *n*-butane it requires about $4\,\text{kcal}\,\text{mol}^{-1}$ ($17\,\text{kJ}\,\text{mol}^{-1}$) to change from the *trans* to the eclipsed positions (Figure 3.15). This accounts for the slight, but significant, buckling (the Pitzer strain) of 5-membered rings. The *skew* or *gauche* arrangement is much more stable, depending on the size of the groups attached to the bond. If these are very small, its stability approaches that of a *trans* arrangement. The

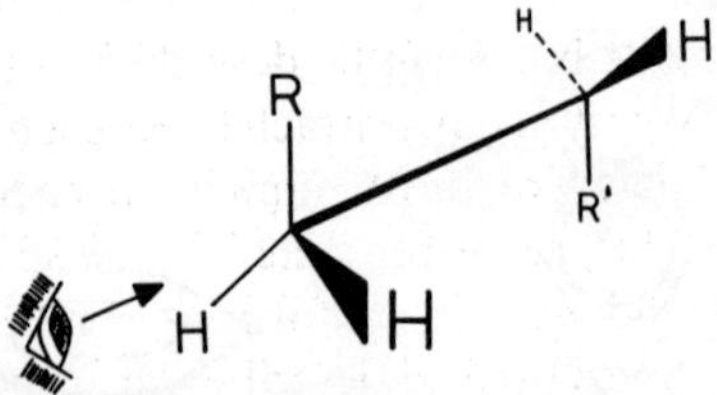

FIGURE 3.14 Representation of conformation by Newman projections; these indicate the arrangement when the bond is viewed end-on. For showing the torsion angle, the bond between the nearer atom and the rest of the molecule is set with the largest groups at 12 o'clock; the above *trans* conformation ($\tau = 180°$) is therefore represented:

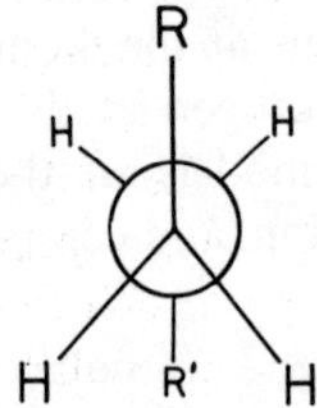

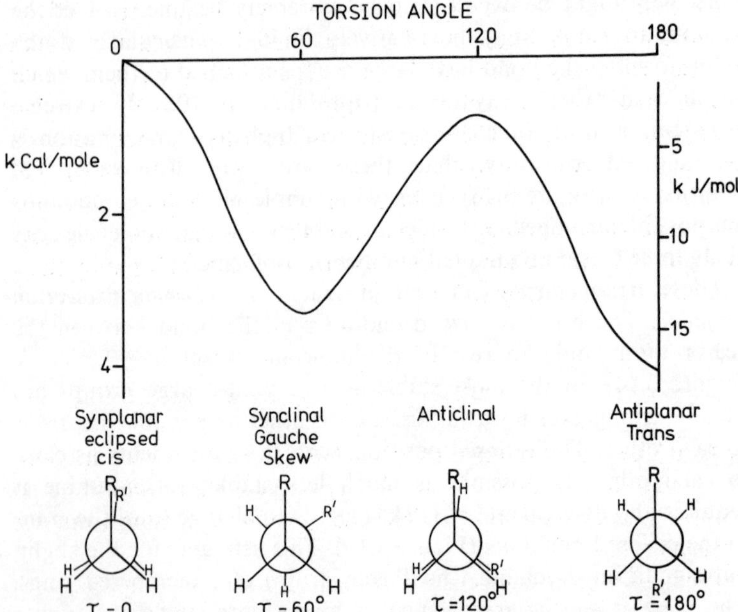

FIGURE 3.15 The relation between energy and torsion angle with the names given to particular conformations. The values are for the conformers of *n*-butane.

rotation about a single bond can be described in terms of the angle between the two largest groups attached to each atom, called the *torsion angle*, τ. There are three minima in energy for values of $\tau = 60°$, $180°$ and $240°$, corresponding to *gauche* (skew, synclinal), *trans* (antiplanar) and *gauche* conformers respectively. The word 'synplanar' is sometimes used to describe the *cis* conformer and the word 'anticlinal' to describe the conformer with $\tau = 120°$; these correspond to the unstable arrangements.

Torsion angles and bond lengths together can be used to describe the relative position of atoms in a molecule. For example, the relative positions of two atoms X and Y at each end of a single bond can be described by the bond lengths, X—C, C—C and C—Y and the two torsion angles —X—C—C— (τ_1) and —C—C—Y— (τ_2). If the stability of the various arrangements can be measured or calculated, it can be represented as a map showing energy contours for different values of τ_1 and τ_2. Such calculations can be made by the theoretical chemist from molecular orbital theory[17] and the preferred conformations correspond to the values of τ_1 and τ_2 which give the energy minima.

An example is shown in Figure 3.16 in which the energy of the benzoylcholine molecule has been calculated for different values of the torsion angles about the bond between the ether oxygen and the (β) methylene group, τ_1, and between the two methylene groups, τ_2. The preferred conformations about the bond between the two methylene groups are *trans* ($\tau_2 = 180°$) and modified *gauche* ($\tau_2 = 80°$). The preferred conformations about the ether-methylene bond are complicated because the oxygen is only divalent and the two lone pairs of electrons will be affected by the π electrons in the conjugated

$$\begin{array}{c} \text{O} \\ \| \\ \text{Ph—C—} \end{array}$$

system. The energy minima indicate that two *gauche* conformations are preferred and also a form with $\tau_1 = 120°$, which would be an *eclipsed* form, except that the eclipsed group is missing from the oxygen atom.

Predictions from molecular orbital theory can be compared with structures derived from X-ray crystallography but the structure in the crystal depends on the packing of the molecules and there may be intermolecular or intramolecular forces which stabilise a

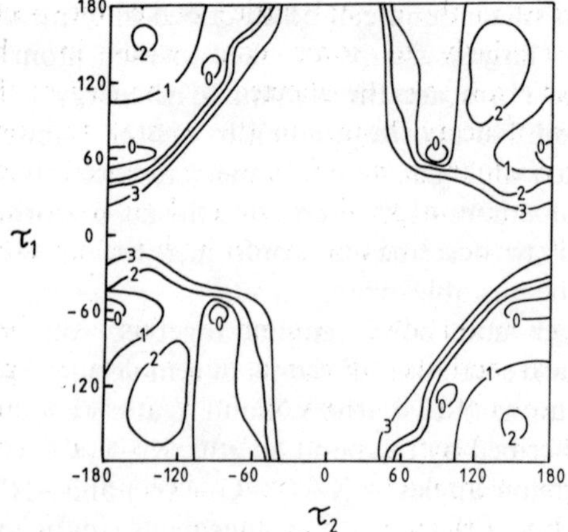

FIGURE 3.16 The energy of the benzoylcholine molecule calculated from molecular orbital theory for various values of the torsion angles, τ_1 and τ_2. (Redrawn from results of Courbeils and Pullman, *Mol. Pharmacol.* (1972), no. 8, p. 281.)

There are six minima indicated with values:

τ_1	τ_2
60	80
60	180
-60	-80
-60	180
120	-80*
-120	80

and the conformer marked with an asterisk can be represented:

$\tau_1 = 120°$ $\tau_2 = -80°$

particular conformer in the crystal lattice in an unexpected way. This applies particularly to ionic compounds in which the conformation of one ion may be affected by the size of the other. The conformation of acetylcholine in the crystal, for instance, is different in the bromide from what it is in the chloride. Accordingly the preferred conformation predicted by molecular orbital theory cannot always be checked by comparison with the structure in the crystal.

The same complication affects attempts which have been made to relate biological activity to the arrangements of groups as assessed from bond lengths and torsion angles observed in crystals. Many studies have been made with compounds related to acetylcholine,[18] but there is always some uncertainty whether these are the same as the preferred conformations in solution. There is also the further uncertainty whether the preferred conformation is necessarily the form which is active at the receptor.

Although the theoretical chemist can now make allowance for the effects of water in his predictions of preferred conformations, it is often possible (and easier) to make a direct experimental assessment of preferred conformations from the nuclear magnetic resonance spectra.

n.m.r. Spectra and Preferred Conformations. If atoms whose nuclei have a magnetic moment, such as ^{1}H, ^{13}C, ^{19}F or ^{31}P, are placed in a strong magnetic field, they may absorb energy of a suitable wavelength, which is in the radio frequency range. A *nuclear magnetic resonance* spectrometer therefore consists of a powerful magnet with a uniform field in which the sample is placed, a radio transmitter supplying energy and a receiver for detecting which frequencies have been absorbed. Scanning can be achieved by altering either the radio frequency or the magnetic field. For the proton and for ^{19}F and ^{31}P only two orientations are possible in the magnetic field and the energy absorbed by a proton in going from the lower energy state to the higher is produced by frequencies of about 60 MHz if the field strength is about 1.4 tesla. The exact frequency absorbed is slightly different for different protons, depending on their own magnetic environment in the molecule.

Absorption occurs only over a very narrow range, usually within 600 Hz in 60 MHz (10 parts per million) but with higher field strengths and frequencies there is a bigger absolute separation, i.e.

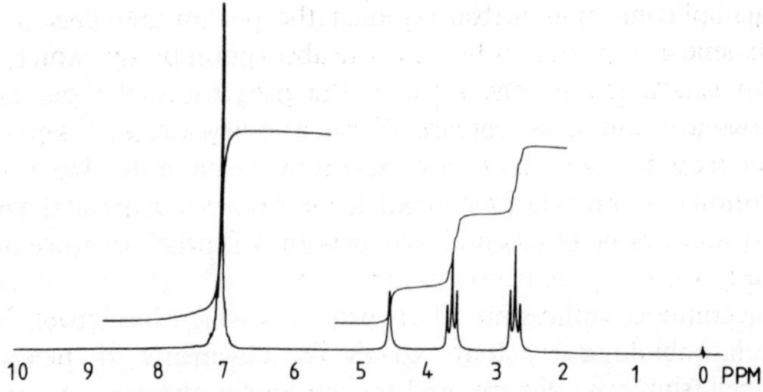

FIGURE 3.17 Proton magnetic resonance spectrum (100 MHz) of $PhCH_2CH_2OH$ dissolved in carbon tetrachloride. The chemical shifts are expressed in parts per million, relative to tetramethylsilane. The upper trace shows the integral. Note the similar splitting of the two triplets, indicating the coupling of the protons in the two methylene groups. The lines in both sets of triplets are separated by about 8 Hz.

the absorption peaks are further apart in hertz, though the same in parts per million. The position of an absorption peak is conveniently expressed relative to that of an internal standard, which is usually tetramethylsilane (Me_4Si; TMS), which absorbs at a higher field than most carbon compounds containing hydrogen. The position of the absorption maximum depends on the electronegativity of the atoms to which the proton is attached and moves to lower fields, for example, with protons which have an oxygen atom in the vicinity (oxygen is more electronegative than carbon, which is slightly more electronegative than silicon). This effect of atoms and groups on the absorption is known as the chemical shift and typical values can be found in tables.[19]

The n.m.r. spectrum of 2-phenylethanol ($PhCH_2CH_2OH$) dissolved in carbon tetrachloride is shown as an example (Figure 3.17). There are four absorption peaks but two are split into triplets. The five protons attached to the benzene ring absorb furthest 'downfield' and the separation from the TMS absorption, δ, is 7.1 parts per million (710 Hz in 100 MHz). The proton attached to the hydroxyl group absorbs further upfield ($\delta = 4.6$) and the triplets correspond to the protons in the methylene groups attached to the oxygen ($\delta = 3.6$) and to the benzene ring ($\delta = 2.7$). The number of protons absorbing can be assessed from the integral of the absorption peak, which has also been recorded and the trace shows that the integrals are in the ratio of 5:1:2:2.

The splitting of the absorption of the protons attached to the methylene groups occurs because the absorption of a proton in one group is affected by the state of the protons in the other; the protons are said to be *coupled*. There are only two states possible for a proton and if these are represented α and β, the possible conditions of the two protons in a methylene group are: both α, one α and one β, one β and one α, both β. The absorption of the protons on one methylene group is therefore split into a triplet, corresponding to the state of the protons in the other group being both α, one α and one β or both β. The absorption of the second group is similarly affected by the state of the protons in the first group and the coupling of protons in the two groups is indicated by the similarity observed in the splitting. The lines in the triplets are separated by the same amount, about 8 Hz, in both sets. This separation represents the coupling constant, J, and its size is independent of the field strength.

The coupling constant depends upon the torsion angle between the two groups. It is very different, for example, if the methylene protons are *eclipsed* rather than *gauche* and different again if they are *trans*. The variation of J with the

$$\begin{array}{ccc} \text{H} & & \text{H}' \\ \diagdown & & \diagup \\ & \text{C—C} & \end{array}$$

torsion angle is described by the Karplus relation (Figure 3.18).

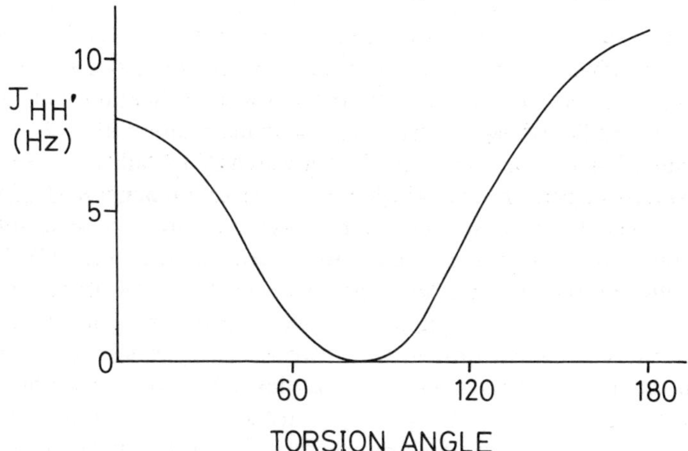

FIGURE 3.18 The Karplus relation between the vicinal coupling constant, $J_{HH'}$, and the torsion (or 'dihedral') angle.

For the *trans* conformer J should be about 11, whereas for the *gauche* conformer it should be about 2. The value $J = 8$ therefore suggests that there is a mixture of *trans* conformer with some *gauche*, unless the molecules are entirely in the eclipsed form ($J = 8$ for torsion angles of $0°$ and $120°$), which is most unlikely. It is possible to calculate the proportion of conformers, therefore, from coupling constants observed in n.m.r. spectra, though the details are usually much more complex than has been indicated in this over-simplified account.

Partington, Feeney and Burgen,[20] for example, have calculated the proportion of *gauche* and *trans* conformers present in many compounds related to acetylcholine from their n.m.r. spectra in D_2O. In choline the proportion of *gauche* conformer was 89%, in acetylcholine it was apparently 100%, whereas in acetylthiocholine it was only 16%. The high proportion of *gauche* conformer indicates an interaction between the ether oxygen $(-O-)$ and the positively charged trimethylammonium group $(-\overset{+}{N}Me_3)$, which is also seen in the crystal. The proportions of the conformers present make it possible to estimate the differences in their free energy from the van't Hoff relation. For example, the proportion of *gauche* and *trans* conformers is at least $20:1$ (allowing for a 5% error in the estimate that it is 100% *gauche*) so it would require at least $1.4 \times \log 20 = 1.8$ kcal mol^{-1} (7.6 kJ mol^{-1}) to convert the *gauche* conformer into the *trans*. For choline the value is $1.4 \times \log 8 = 1.3$ kcal mol^{-1} (5.3 kJ mol^{-1}).

The n.m.r. spectra therefore afford important information about the shapes and preferred conformations of drugs and the energy needed to change from one shape to another. This may help to answer the question 'is the preferred conformation in solution necessarily the one which is biologically active?' by indicating what other conformations are energetically possible: n.m.r. techniques are also of immense importance in studying drug-receptor interactions because the signals obtained from the drug will be modified by the presence of the protons in the receptor with which it is interacting. There are often practical problems, such as the absorption from protons in water itself, and the whole spectrum may be extremely complex, but it may be possible to obtain information from studying only part of the spectrum. The absorptions of the six histidine residues in the enzyme dihydro-folate reductase, for example, are identifiable and three of these are shifted by the presence of inhibitors, such as trimethoprim.[21] By examining the effects of different ligands it is possible to see that only one of these histidine residues is probably involved in the binding of the glutamic acid part of dihydrofolate.

This extremely brief introduction is intended simply to point out to the biologist the great relevance of this branch of physical chemistry and to stimulate him to read further.[22] Nuclear magnetic resonance has the advantage over a related physical technique, electron spin resonance (e.s.r.), in that it studies the actual molecules which are of interest, not related compounds with the spin-label necessary for e.s.r. These spin-labelled compounds are usually obtained by incorporating an *N*-oxide group (which has an unpaired electron) into the molecule and this is likely seriously to affect its biological properties. The limitations imposed on n.m.r. are mainly the need to have the binding protein pure and preferably in solution, though some studies involving solid fragments have been made. With a subject which is developing so rapidly, however, it is difficult to see what the limits are to future progress.

References

1. O. Exner, *Coll. Czech. Chem. Comm.* (1967), no. 31, p. 7.

2. F. J. Millero, *Chem. Review* (1971), no. 71, p. 147; R. B. Barlow, B. M. Lowe, J. D. M. Pearson, H. M. Rendall and G. M. Thompson, *Mol. Pharmacol.* (1971), no. 7, p. 357; R. B. Barlow and F. M. Franks, *British Journal Pharmacol.* (1973), no. 49, p. 480.

3. L. E. Sutton, *Interatomic Distances* and Supplement (Chemical Society special publications nos. 11 and 18, London, 1958 and 1965); A. S. Dreiding, *Helv. Chim Acta* (1959), no. 42, p. 1339.

4. R. B. Barlow, F. M. Franks and J. D. M. Pearson, *Journal Pharm. Pharmacol.* (1972), no. 24, p. 753.

5. P. S. Portoghese, *Annual Review Pharmacol.* (1970), no. 10, p. 51.

6. K. Frendenberg, *Stereochimie* (Deuticke, Leipzig, 1933), p. 672–720; these formulae for (+)- and (−)-glyceraldehyde are however based on the formulae for (+)- and (−)-tartaric acid assumed by E. Fischer, *Ber. Deut. Chem. Ges.* (1891), no. 24, p. 2683.

7. K. Freudenberg and F. Rhino, *Ber. Deut. Chem. Ges.* (1924), no. 57, p. 1547; K. Frendenberg and M. Meister, *Liebig's Ann.* (1935), no. 518, p. 86.

8. J. C. Craig and S. K. Roy, *Tetrahedron* (1965), no. 21, p. 401.

9. A. R. Battersby, I. R. C. Bick, W. Klyne, J. P. Jennings, P. M. Scopes and M. J. Vernengo, *Journal Chem. Soc.* (1965), p. 2239.

10. J. M. Bijvoet, A. F. Peerdeman and A. J. van Bommel, *Nature* (1951), no. 168, p. 271.

11. R. S. Cahn, C. K. Ingold and V. Prelog, *Experientia* (1956), no. 12, p. 81.

12. R. S. Cahn, C. K. Ingold and V. Prelog, *Angew. Chimie* (1966), no. 5, p. 385.

13. D. T. Elmore, *Peptides and Proteins* (Cambridge University Press, Cambridge, 1968), p. 63.

14. J. E. Blackwood, C. L. Gladys, K. L. Loening, A. E. Petrarca and J. E. Rush, *Journal American Chem. Soc.* (1968), no. 90, p. 509.

15. G. Eglinton, in *Physical Methods in Organic Chemistry*, J. C. P. Schwarz (ed.) (Oliver and Boyd, Edinburgh, 1964); L. J. Bellamy, *The Infrared Spectra of Complex Molecules*, 2nd edition (London, 1958); C. J. Pouchert, *The Aldrich Library of Infrared Spectra* (The Aldrich Chemical Co., USA, 1970); A. J. Gordon and R. A. Ford, *The Chemist's Companion* (Wiley, New York, 1972), p. 168.

16. W. H. Pirkle, *Journal American Chem. Soc.* (1966), no. 88, p. 1837; M. Raban and K. Mislow, *Tetrahedron Letters* (1966), p. 3961; J. A. Dale, D. L. Dull and H. S. Mosher, *Journal Org. Chem.* (1969), no. 34, p. 2543; H. L. Goering, J. N. Eikenberry and G. S. Koermer, *Journal American Chem. Soc.* (1971), no. 93, p. 5913.

17. B. Pullman and A. Pullman, *Quantum Biochemistry* (Interscience, New York, 1963); L. B. Kier, *Molecular Orbital Studies in Chemical Pharmacology* (1970); W. G. Richards, *Quantum Pharmacology* (Butterworth, London, 1978).

18. R. W. Baker, C. H. Chothia, P. Pauling and T. J. Petcher, *Nature* (1971), no. 230, p. 439; A. F. Casy, in *Progress in Medicinal Chemistry* 11, G. P. Ellis and G. B. West (eds.) (North-Holland, Amsterdam, 1975); A. F. Casy, *Ann. Rep. Chem. Soc. B* (1975), p. 477.

19. A. J. Gordon and R. J. Ford, *The Chemist's Companion* (Wiley, New York, 1972), p. 252.

20. P. Partington, J. Feeney and A. S. V. Burgen, *Mol. Pharmacol.* (1972), no. 8, p. 269.

21. G. C. K. Roberts, in *Drug Action at the Molecular Level* (Macmillan, London, 1977), p. 127; A. S. V. Burgen, in *Biological Activity and Chemical Structure*, edited by J. A. Keverling-Buisman (Elsevier, Amsterdam, 1977), p. 83.

22. J. A. Pople, W. G. Schneider and H. J. Bernstein, *High-resolution Nuclear Magnetic Resonance* (McGraw-Hill, New York, 1959); D. W. Mathieson, *Nuclear Magnetic Resonance for Organic Chemists* (Academic Press, London, 1967); A. F. Casy, *PMR Spectroscopy in Medicinal and Biological Chemistry* (Academic Press, London, 1971).

CHAPTER 4

Chemical Processes where Size and Shape are Important

Adsorption

Freundlich Isotherm

It has long been known that molecules may become concentrated, i.e. adsorbed, at the surfaces of some solids. This is of practical value, for instance, when charcoal is used to remove traces of coloured impurities from solutions or to adsorb vapours in domestic cooker-hoods and in gas masks. It might be expected that the amount adsorbed for a given weight of adsorbing material would depend on the partial pressure or concentration of the adsorbed species and on its nature, thus:

$$x = KP^n$$

where x is the amount adsorbed for unit mass of adsorbent, P is the partial pressure of the adsorbed material (or concentration if it is in solution) and K and n are constants. This expression, known as the *Freundlich adsorption isotherm*, was originally arrived at empirically and has been found to fit many experimental results. It can be written:

$$\log x = \log K + n \log P$$

so the graph of $\log x$ against $\log P$ should be a straight line with a slope of n and an intercept of $-\log K/n$ when $\log x = 0$. There should be no upper limit to the amount which can be adsorbed and as the partial pressure is increased it should eventually reach the saturated vapour pressure and condensation will occur.

For the adsorption of vapours on charcoal there is a good correlation between the amount of material adsorbed and its critical temperature and the compounds which are most easily liquefied are those found to be most strongly adsorbed. The forces between the adsorbed molecules and the charcoal thus appear to be similar to those between molecules in the liquid state. The process is described as physical or van der Waals' adsorption and the enthalpy of adsorption is often comparable with the latent heat of evaporation (usually less than $10\,\text{kcal mol}^{-1}$ ($42\,\text{kJ mol}^{-1}$)). The large size of the amounts adsorbed by charcoal, then, can be ascribed to its ability to allow the molecules to become organised in a manner similar to the liquid and also to the physical state of the charcoal, which is finely divided or porous and has a very large surface area.

Langmuir Isotherm

If the process of adsorption is regarded as occurring only at certain sites, it can be represented:

$$X+S \rightleftharpoons XS$$

The rate of adsorption will be $k_{+1}Pn(1-\theta)$, where k_{+1} is a rate constant, P is the partial pressure (or concentration), n is the number of sites in unit mass of adsorbent and θ is the proportion of these which is occupied. The rate of desorption will be $k_{-1}n\theta$ and at equilibrium these two rates will be equal so

$$\frac{k_{+1}}{k_{-1}}P = \frac{\theta}{1-\theta}$$

and $\theta = bP/(1+bP)$, where b, the adsorption coefficient $= k_{+1}/k_{-1}$.

When P is large, $bP/(1+bP) \rightarrow 1$ but when P is small, $1+bP \doteqdot 1$, so $\theta = bP$. The amount of material, x, adsorbed by unit mass of adsorbent is directly related to θ, the proportion of sites occupied; in fact $x = n\theta(M/N)$, where M is the molecular weight of the adsorbed molecules and N is Avogadro's number. At low partial pressures (or concentrations), therefore, adsorption will apparently be a first order process and directly related to partial pressure, but as this rises the process becomes zero order, with the amount adsorbed independent of the pressure as the adsorption sites become saturated.

This isotherm relating partial pressure and amount adsorbed is one of a number of expressions derived by Langmuir[1] and can be extended to the binding of molecules with many biological materials such as the binding of compounds to plasma proteins, of substrates to enzymes and of drugs to receptors.

The *adsorption coefficient, b,* and the Gibbs free energy of adsorption per mole, ΔG, are connected by the van't Hoff relation:

$$-\Delta G = RT \ln b$$

The value of ΔG will be determined by the strength of the forces between the adsorbed molecules and binding site. These are discussed later (page 181) but binding is usually the result of a combination of weak chemical forces, so if it is to be appreciable, there must be a considerable area in the binding site where interactions with an adsorbed molecule are possible. Strong binding is likely to be associated with a highly 'structured' site and a fit between many groups in the site and the adsorbed molecules.

This type of adsorption is termed *chemisorption* to distinguish it from physical adsorption and is usually associated with a relatively high enthalpy of adsorption, ΔH. The process occurs spontaneously, so ΔG is negative and, if the order in the system is increased by the process of adsorption, ΔS is also negative. Because $\Delta G = \Delta H - T\Delta S$, i.e. $\Delta H = \Delta G + T\Delta S$ the value of ΔH will be even more negative than ΔG and the process should be markedly exothermic. With adsorption from aqueous solution, however, interactions between the molecules and water must be considered and the order in the system need not necessarily increase.

Analysis of Results

The amount of material, x, adsorbed by unit mass of adsorbent is $kbP/(1+bP)$ where $k = nM/N$, and this equation, which represents a hyperbola, can be rewritten in three different ways which transform it into a straight line:

$$\frac{1}{x} = \frac{1}{kbP} + \frac{1}{k}$$

or

$$\frac{x}{P} = kb - bx$$

or

$$\frac{P}{x} = \frac{1}{kb} + \frac{P}{k}.$$

Accordingly:

(i) if $1/x$ is plotted against $1/P$, a straight line should be obtained which has a slope of $1/kb$ and when

$$\frac{1}{x} = 0, \qquad \frac{1}{P} = -b;$$

(ii) if x/P is plotted against x, a straight line should be obtained which has a slope of $-b$ and when

$$\frac{x}{P} = 0, \qquad x = k;$$

(iii) if P/x is plotted against P, a straight line should be obtained which has a slope of $1/k$ and when

$$\frac{P}{x} = 0, \qquad P = -\frac{1}{b}.$$

Because $bP/(1+bP) = 1$ when P is large, k corresponds to x_{max}, the maximum amount of material which can be absorbed. If $P_{1/2}$ is the partial pressure which produces the adsorption of half the maximum amount, $x = k/2$ and the equation becomes:

$$\frac{k}{2}(1 + bP_{1/2}) = kbP_{1/2}$$

so $bP_{1/2} = 1$, and the adsorption coefficient, b, is the reciprocal of the partial pressure (or concentration) which produces half-saturation.

The equation can also be written:

$$bP = \frac{x}{k-x}$$

so if x_{max} is known it is possible to plot $x/(x_{max}-x)$ against P and a straight line should be obtained which has a slope of b. If $\log[x/(x_{max}-x)]$ is plotted against $\log P$ a straight line should be obtained and when $\log[x/(x_{max}-x)] = 0$, $\log P = -\log b$. The slope of the line should be 1. The expression $\log[x/(x_{max}-x)]$ is logit $[x/x_{max}]$; logit (A) is $\log[A/(1-A)]$ so logit (x/x_{max}) is:

$$\log\left[\left(\frac{x}{x_{max}}\right)\Big/\left(1 - \frac{x}{x_{max}}\right)\right] = \log\left(\frac{x}{x_{max}-x}\right)$$

This method of analysis was originally applied by Hill[2] to the reaction between oxygen and haemoglobin, which was written:

$$(Hb)_n + nO_2 \rightleftharpoons (HbO_2)_n$$

If y is the proportion of oxygenated haemoglobin $(HbO_2)_n$, there will be a proportion $(1-y)$ available for reaction with oxygen so the rate of formation will be $k_{+1}(1-y)[O_2]^n$. The rate of breakdown will be $k_{-1}y$ and if K is the association constant (k_{+1}/k_{-1}), then

$$K[O_2]^n = \frac{y}{1-y}.$$

The fraction $y/(1-y)$ has been called the degree of saturation and the concentration of oxygen $[O_2]$ is indicated by its partial pressure, P, so the equation becomes:

$$KP^n = \frac{y}{1-y} \quad \text{or} \quad y = \frac{KP^n}{1+KP^n}.$$

The graph of logit (y) against $\log P$ should therefore be a straight line with a slope of n and when logit $(y) = 0$, $\log P = -(\log K)/n$. The 'Hill plot' is therefore particularly useful because it can be applied in situations where n is not limited to 1. It can also be used to see when molecules are acting cooperatively, with the adsorption of one molecule facilitating the adsorption of another (or making it

TABLE 4.1: *The partial pressure of ethyl chloride and the mass adsorbed on 10 g charcoal at* $-15.3°C$: x_{calc} *is the value obtained from the hyperbola* $x = 4.77P/(2.56+P)$ *(page 124)*

Partial pressure (P) (mm Hg)	Mass adsorbed g/10 g charcoal (x)	x_{calc}	Difference
227.9	5.062	4.717	0.345
195.4	5.012	4.708	0.304
131.1	4.907	4.678	0.229
105.7	4.848	4.657	0.191
58.4	4.657	4.570	0.087
35.6	4.329	4.450	-0.121
26.3	4.098	4.347	-0.249
17.81	3.782	4.170	-0.388
12.02	3.458	3.932	-0.474
5.50	2.845	3.255	-0.410
3.04	2.417	2.589	-0.172
1.41	1.922	1.694	0.228
0.571	1.455	0.870	0.585
0.289	1.168	0.484	0.684
0.119	0.874	0.212	0.662
0.0479	0.661	0.088	0.573
0.0372	0.605	0.068	0.537
0.0177	0.465	0.033	0.432

(From results of Goldman and Polanyi, *Z. Phys. Chem.* (1928), no. 132, p. 334.)

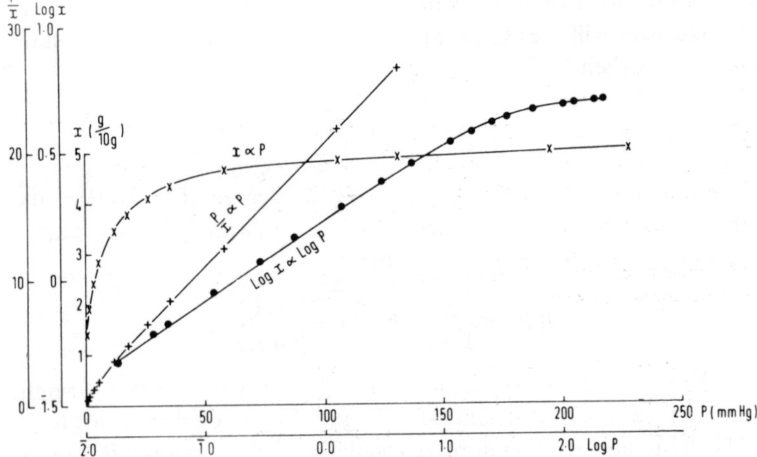

FIGURE 4.1 Adsorption of ethyl chloride on charcoal. The points, ×, show the results in Table 4.1 with the amount adsorbed by 10 g charcoal (x) plotted against the partial pressure (P). The points ● show the graph of $\log x$ against $\log P$; the points + show the graph of P/x against P.

more difficult); n and K will alter with P and the graph is not a straight line. This method, however, requires an accurate estimate of the maximum adsorption (x_{max}).

For example, the mass of ethyl chloride adsorbed on 10 g charcoal at various partial pressures is shown in Table 4.1. The graph of x against P suggests that the process is saturable (Figure 4.1) and departs from the Freundlich isotherm at high partial pressures. The graph of P/x against P suggests that the results are better fitted to the Langmuir adsorption isotherm but it is difficult to be sure about this, because P is plotted on an arithmetic scale so the values for lower partial pressures are compressed together in an unsatisfactory manner; this illustrates the advantages of using a logarithmic scale (but $\log(P/x)$ is not linearly related to $\log P$).

The transformation of the hyperbolic relation between x and P into a straight line should make it easier to see whether the results fit the Langmuir isotherm, because the eye is better able to judge the straightness of a line than the hyperbolic nature of a curve. It is also then relatively easy to calculate the constants k and b which determine the adsorption process. There are, however, complications with the use of the linear transformations, associated with the fact that there are errors attached to the results.

When a straight line is drawn by eye, the ruler is placed so that, as far as seems possible, the points lying off the line are equally distributed on either side. Different people will draw different lines. The position of the line $y = mx + c$ for values of x and y can, however, be located by the *method of least-squares*, which determines the values of m and c such that the sum of $(y_{observed} - y_{calculated})^2$ for all the results is a minimum; it is also possible to estimate the standard errors of m and c (see Appendix).

This is a least-squares fit of y on x, and assumes that the errors in x are negligible compared with those in y. In an adsorption experiment the errors in the amount adsorbed are likely to be much greater than those in the partial pressure or concentration, so the method of least-squares should be applied to the fit of $1/x$ on $1/P$ or to the fit of P/x on P. There will be problems with the fit of x/P on x, though the errors in x/P should be bigger than the errors in x alone.

The errors attached to results may not be constant and if this is so the results should be weighted, so that the more inaccurate results do not have an undue effect. The least-squares fit then determines the values of m and c such that $Sw(y_{obs} - y_{calc})^2$ is a minimum, where w is the weight for each result. Sometimes it may be necessary to assign an arbitrary value to w, but if each value of y is the mean of a number of estimates, the reciprocal of the variance of these estimates is a very suitable weighting factor, i.e. $w = 1/V$ where V is the variance. The fit of the results to the line can be assessed from the correlation coefficient and the residual variance (Appendix) but in most experiments the aim is not merely to test the fit of the Langmuir isotherm but also to measure the constants k and b, which determine the binding. If these are calculated from m or c, it is possible to estimate their standard errors. For this reason the graph of x/P against x is preferable to the other two transformations for estimating the absorption coefficient, b. With the two other transformations the calculation of b involves the intercept on the x-axis and the calculation of the standard error is difficult.

The commercial development of electronic calculators has now made it easier to fit results to a straight-line by the method of least-squares than to draw a graph and fit results by eye. Further, it is also possible to perform a least-squares fit to many other relations, including the hyperbolic expression itself (Appendix). Accordingly, there is no need for any linear transformation; values of x can be fitted by least-squares to values of P and the values of k and b calculated. To do this the equation is often rewritten:

$$x = \frac{MP}{P + K_d}$$

where M is the maximum amount adsorbed and K_d is the dissociation constant, $1/b$.

Whatever procedure is used for calculating k and b, however, it is important to compare the observed and calculated values carefully, to see whether there are systematic differences. The results in Table 4.1, for instance, do not really fit the Langmuir isotherm. The least-squares fit of $1/x$ on $1/P$ indicates saturation with $x = 2.62\,\text{g}/10\,\text{g}$ and for half-saturation $P = 0.098\,\text{mm Hg}$; the fit for x/P on x indicates saturation with $x = 4.24\,\text{g}/10\,\text{g}$ and for half-saturation $P = 0.32\,\text{mm Hg}$; the direct fit to the hyperbola indicates saturation with $x = 4.77\,\text{g}/10\,\text{g}$ and for half-saturation $P = 2.56\,\text{mm Hg}$. With the direct fit to the hyperbola it can be seen that the curve has the wrong shape (Table 4.1) and the estimated maximum is less than some of the experimental values, so it is not possible to make a Hill plot of the results.

The underlying assumption that the adsorption of ethyl chloride on charcoal can be represented as:

$$X + S \rightleftharpoons XS$$

is incorrect. It is, in fact, unlikely that the binding sites on the charcoal are so similar that the free energy of adsorption is the same at all of them and it is also unlikely that with a volatile liquid not far from its boiling point (13°C) only one molecule is adsorbed at each site.

It is quite possible, however, that this equation represents the binding of molecules to specific structures such as enzymes or receptors and even to the binding of drugs to plasma proteins.

The Binding of Drugs to Proteins

Many drugs become bound to plasma proteins and for some it may be necessary to give a large 'loading' dose to fill this compartment (Figure 2.5) before a significant blood-level can be produced. It is usual to express the binding as the ratio, r, of the number of molecules of drug bound to each molecule of protein. If n is the number of binding sites in each molecule of protein,

$$r = n\theta = \frac{nKC}{1+KC}$$

where C is the concentration of the drug and K is its affinity constant (corresponding to the adsorption coefficient, b). Often the dissociation constant, $K_D = 1/K$, is used instead of K so,

$$r = \frac{nC}{K_D + C}.$$

The hyperbolic relationship can be transformed into a linear one by any of the methods described on page 119 but results are often analysed by (ii), in the manner suggested by Scatchard.[3] The equation can be rearranged:

$$r + rKC = nKC \quad \text{so} \quad \frac{r}{C} = nK - rK$$

and the graph of r/C against r (bound/free against bound) should be a straight line with a slope of $-K$ and when $r/C = 0$, $r = n$. If K_D is used instead of K the slope is $-(1/K_D)$.

For example, the binding of benzylpenicillin to bovine serum albumin was studied by Klotz, Urquhart and Weber[4] in experiments in which 10 ml of a buffered solution of albumin (0.408 mM), enclosed in a dialysis bag, were left to equilibrate with 20 ml of a solution of benzylpenicillin in the same buffer. A control experiment was made with the same solution of benzylpenicillin and buffer, but no albumin, in the dialysis bag, and the amount of

TABLE 4.2: *Binding of benzylpenicillin to bovine serum albumin at 5°C*

Free benzyl-penicillin (C)	Moles bound/ mole protein (r)	r/c
31 μM	0.18	5.81×10^3 l mol^{-1}
43	0.23	5.35
116	0.46	3.97
143	0.56	3.92
232	0.85	3.66
309	1.04	3.37
522	1.40	2.68
564	1.52	2.69

(From results of Klotz, Urquhart and Weber, *Arch. Biochem.* (1950), no. 26, p. 427.)

penicillin bound was calculated from the difference between the final concentrations in the two experiments. Some results are shown in Table 4.2; the first value indicates that the final concentration in the presence of the albumin was 31 µM, compared with 55.5 µM in the control without the albumin (not shown). The amount bound is given by the difference in concentration multiplied by the total volume, i.e.

$$(55.5\text{--}31) \times \frac{30}{1000}\,\mu\text{mol} = 0.735\,\mu\text{mol}.$$

There are 4.08 µmol of albumin present so

$$r = \frac{0.735}{4.08} = 0.18.$$

The graph of r/C against r is not a particularly satisfactory straight line (Figure 4.2.) because the two points corresponding to the two lowest values of r lie well above it. These were obtained with the two lowest concentrations, for which the errors are likely

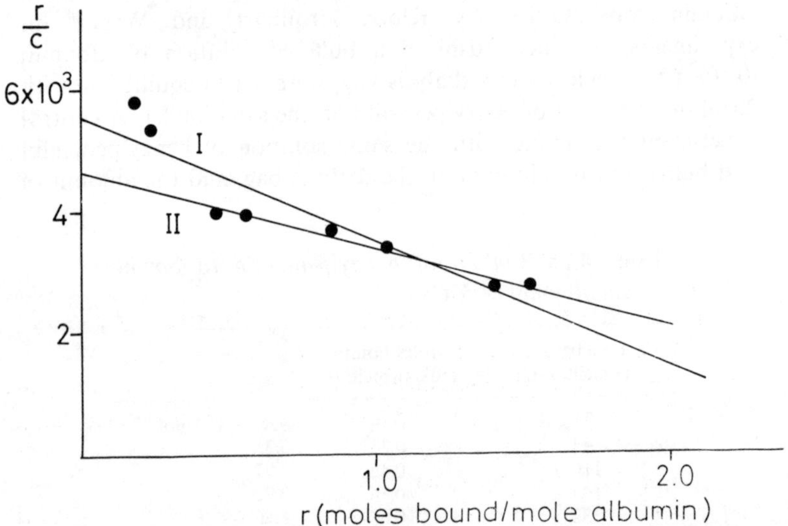

FIGURE 4.2 Binding of benzylpenicillin to bovine serum albumin. The points show a Scatchard plot of the results in Table 4.2, with the number of moles bound/mole of protein (r) divided by the concentration of free benzylpenicillin (C), i.e. r/C, plotted against r. Line I represents the least-squares fit to a straight line for all 8 points and line II for 6 points, omitting the results with the two lowest concentrations which obviously lie off a straight line.

to be greatest. A least-squares fit for all 8 points indicates that $K = 2.08 \times 10^3 \, \mathrm{l\,mol^{-1}}$ and there are 2.7 binding sites per molecule of bovine serum albumin. If only the 6 points lying more obviously on the line are taken, $K = 1.32 \times 10^3$ and $n = 3.5$ binding sites/molecule protein. If the results are fitted directly to a hyperbola the estimate of K is 1.51×10^3, or 1.42×10^3 if only 6 points are taken. The method of analysing the results, therefore, affects the estimates obtained, particularly when the results have large errors or do not really fit the equation which is supposed to describe the binding process.

TABLE 4.3: *Binding of labelled acetylcholine to a phospholipid*

Acetylcholine concentration (C)	Radioactivity bound to phospholipid dpm/μg	Amount acetylcholine bound/μg (r)	$r/C*$
0.056 μM	176	0.85×10^{-11}	15.2×10^{-5}
0.22	334	1.62	7.36
1.1	570	2.76	2.51
2.4	880	4.27	1.78
3.8	1465	7.10	1.87
7.3	1750	8.48	1.16
19.0	3000	14.54	0.765
42.0	3890	18.86	0.449

$*\,\mu g \times \mathrm{litres} \times \mathrm{mol}^{-1}$.

(From results of De Robertis, Lunt and La Torre, *Mol. Pharmacol.* (1971), no. 7, p. 100.)

The sensitivity can be enormously improved by working with compounds with a radioactive label and Table 4.3 shows results obtained by De Robertis[5] for the binding of labelled acetylcholine to a phospholipid, which could be receptor material. Aliquots of this material were left to equilibrate with various concentrations of acetylcholine and then separated and the radioactivity counted and expressed as disintegrations/minute for 1 μg phospholipid. The binding is thus known much more accurately than previously because it is measured directly and not calculated by subtraction. It was known that 4950 disintegrations/minute correspond to 2.4×10^{-10} moles of acetylcholine so 176 correspond to 0.85×10^{-11} moles and r/C is 1.52×10^{-4} (Table 4.3), where r is the amount bound by 1 μg (not 1 mole) of phospholipid. A Scatchard plot (Figure 4.3) can be interpreted by supposing that there are two

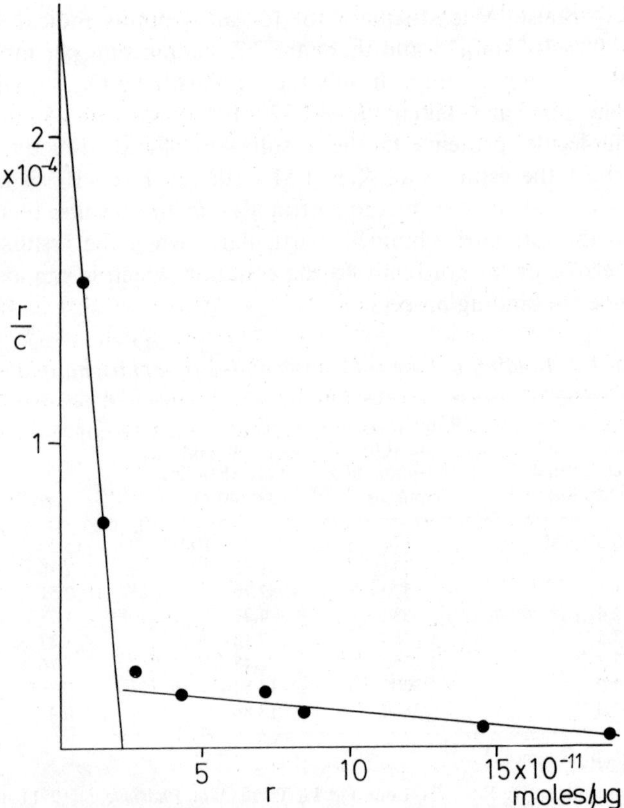

FIGURE 4.3 Binding of acetylcholine to a phospholipid. The points show a Scatchard plot of the results in Table 4.3, with the number of moles of acetylcholine bound/µg phospholipid (r) divided by the concentration of acetylcholine (C), i.e. r/C, plotted against r. There appear to be two binding sites in this material.

binding processes, represented by the two lines, whose slopes correspond to a high affinity binding (at low concentrations) with

$$K_H = \frac{2.1 \times 10^{-4}}{2.0 \times 10^{-11}} = 1.05 \times 10^7,$$

and a low affinity binding (at high concentrations) with

$$K_L = \frac{0.08 \times 10^{-4}}{10.0 \times 10^{-11}} = 8 \times 10^4.$$

If the high affinity binding corresponds to 1 mole of acetylcholine and 1 mole of phospholipid, 1 µg corresponds to 2.3×10^{-11} moles and the molecular weight of the phospholipid is approximately 43,000.

When drugs become bound at more than one site, as in this experiment, the total amount bound/mole of protein,

$$r = \frac{n_1 K_1 C}{1 + K_1 C} + \frac{n_2 K_2 C}{1 + K_2 C} + \ldots$$

where n_1 is the number of binding sites/mole for which the affinity constant of the drug is K_1 and so on. Portions of the Scatchard plot may be linear enough for it to be possible to calculate K and n for each site, as in Figure 4.3; it is also possible to fit the results directly to the whole expression by the method of least-squares, though it may be difficult to ensure that the fit is not determined exclusively by results at one end of the binding curve.

This type of experiment is commonly used to study the binding of drugs to preparations of receptor material, often in the form of membrane fragments, which can be separated after equilibration by centrifugation. It is not necessary to limit the work only to labelled drugs, however, because the binding of unlabelled drugs can be studied by competition with that of a labelled drug. It is therefore necessary to consider next the adsorption of two species to the same receptor.

Adsorption of Two Species

If two different molecules *compete* for the same binding site and the process can be represented:

$$X_1 + S \rightleftharpoons X_1 S$$

$$X_2 + S \rightleftharpoons X_2 S$$

then

$$b_1 P_1 = \frac{\theta_1}{1 - \theta_1 - \theta_2}$$

and

$$b_2 P_2 = \frac{\theta_2}{1 - \theta_1 - \theta_2}$$

where P_1 and P_2 are the partial pressures, or concentrations, of X_1 and X_2, b_1 and b_2 are their adsorption coefficients, and θ_1 and θ_2 are the proportions of the sites occupied (so the proportion unoccupied is $1 - \theta_1 - \theta_2$). Dividing the two,

$$\frac{b_1 P_2}{b_2 P_2} = \frac{\theta_1}{\theta_2}$$

but

$$\theta_2 = b_2 P_2 (1 - \theta_1 - \theta_2) = \frac{(1 - \theta_1) b_2 P_2}{1 + b_2 P_2}$$

so

$$\frac{b_1 P_1}{b_2 P_2} = \frac{\theta_1 (1 + b_2 P_2)}{(1 - \theta_1) b_2 P_2}$$

and

$$b_1 P_1 = \frac{\theta_1}{1 - \theta_1}(1 + b_2 P_2) \quad \text{or} \quad \theta_1 = \frac{b_1 P_1}{1 + b_1 P_1 + b_2 P_2}.$$

These relationships are illustrated in Figure 4.4. Because the two species X_1 and X_2 are competing for the same binding site, an

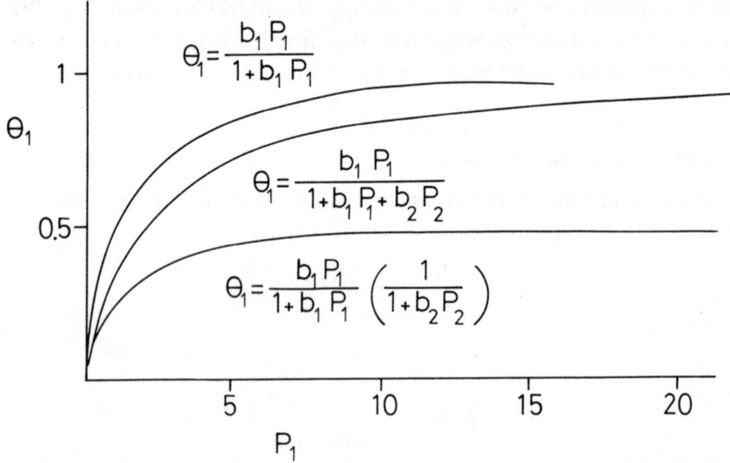

FIGURE 4.4 Adsorption of two species; the proportion of sites (θ_1) occupied by the first species is plotted against the partial pressure (or concentration) of this species for an adsorption coefficient $b_1 = 1$. The middle curve indicates what happens in the presence of a second species which behaves competitively and the lower curve what happens when the second species behaves noncompetitively; for both these curves $b_2 P_2 = 1$. Note that for the competitive situation P_1 must be multiplied by $1 + b_2 P_2$ ($= 2$) to produce the same value of θ_1 as was obtained in the absence of the second species. The antagonist alters the scale of the x-axis. If the second species acts noncompetitively it alters the scale of the y-axis.

increase in the concentration of X_1 will increase the proportion of sites occupied by these molecules even in the presence of the second species. If the proportion of sites occupied by X_1 at a partial pressure, or concentration, P_1 is θ_1 in the absence of the second species, it will be reduced when this is present at a partial pressure or concentration, P_2, but may be restored to θ_1 by increasing P_1 to P'_1. Accordingly

$$b_1 P_1 = \frac{\theta_1}{1 - \theta_1}$$

and

$$b_1 P'_1 = \frac{\theta_1}{1 - \theta_1}(1 + b_2 P_2)$$

and the ratio $P'_1/P_1 = 1 + b_2 P_2$.

There are, however, situations in which the binding of the first species X_1 is affected by the second species, X_2, but where the binding of X_2 is independent of the first species, X_1. This would occur, for instance, if the second species combined very strongly with the binding site. For the binding of the first species the equation will be

$$b_1 P_1 = \frac{\theta_1}{1 - \theta_1 - \theta_2}$$

as before, but for the second species it will be

$$b_2 P_2 = \frac{\theta_2}{1 - \theta_2}$$

so

$$\theta_2 = \frac{b_2 P_2}{1 + b_2 P_2} \quad \text{and} \quad 1 - \theta_2 = \frac{1}{1 + b_2 P_2}.$$

From the first equation,

$$\theta_1 = b_1 P_1 (1 - \theta_1 - \theta_2)$$

$$= \frac{b_1 P_1 (1 - \theta_2)}{1 + b_1 P_1}$$

$$= \frac{b_1 P_1}{1 + b_1 P_1}\left(\frac{1}{1 + b_2 P_2}\right)$$

The proportion of sites occupied by the first species is reduced in the presence of the second species but in the situation just considered, which is described as *noncompetitive*, an increase in the partial pressure, or concentration, of the first species will not properly reverse this (Figure 4.4). The first species can never saturate all the sites; when P_1 is very large and

$$\frac{b_1 P_1}{1 + b_1 P_1} \rightarrow 1$$

the proportion of sites occupied can only be $1/(1 + b_2 P_2)$.

As before (p. 118), the amounts of material, x_1 and x_2, adsorbed by unit mass of adsorbent are directly proportional to the number of sites occupied, so

$$\theta_1 = \frac{x_1}{x_{1\,max}} \quad \text{and} \quad \theta_2 = \frac{x_2}{x_{2\,max}},$$

where $x_{1\,max}$ and $x_{2\,max}$ are the maximum amounts which can be adsorbed (separately). If the two species are competing for the same receptor, $x_{1\,max}$ and $x_{2\,max}$ should be equal if they are measured on a molar basis.

The above equations describe conditions at equilibrium. The rates at which adsorption or desorption occur will be determined by the rate constants, and may possibly be limited by diffusion. The ratio of the forward and backward rate constants determines the equilibrium constant but it is possible to have two adsorption processes with the same equilibrium constant and quite different rate constants. The rate of desorption of X_2, for instance, may be very slow, even though b_2 may be similar to b_1, and if insufficient time is allowed the antagonism may appear to be noncompetitive. As it stands, the model does not suppose that there can be active displacement of one species from the binding site by the other. This would require a step such as:

$$X_1 + X_2 S \rightleftharpoons X_1 S + X_2$$

If this does not occur to any extent, the break-up of $X_1 S$ or $X_2 S$ is determined only by the rate constants (not by P_1 or P_2) and if one of these is very slow the binding of that species may appear to be unaffected by an increase in the concentration of the other. In this situation an antagonism which is competitive will appear to be noncompetitive if measurements are made before equilibrium has been reached.

The interaction of two drugs binding to the same site on plasma proteins can sometimes be therapeutically important. Aspirin and other salicylates bind to the same sites as does the anticoagulant drug, warfarin, so they may produce a dangerous rise in the level of free warfarin if given to patients being treated with it.

As already mentioned, the adsorption of two species is of great importance when the binding of drugs to receptor fragments is studied by competition between one species which is radio-labelled and another which is not. If the fraction of sites occupied by the labelled species is θ_1' in the presence of the second species and θ_1 in its absence,

$$\frac{\theta_1'}{\theta_1} = \frac{1 + b_1 P_1}{1 + b_1 P_1 + b_2 P_2}$$

but

$$1 + b_1 P_1 = 1 + \frac{\theta_1}{1 - \theta_1} = \frac{1}{1 - \theta_1} \quad \text{and} \quad \frac{\theta_1'}{\theta_1} = \frac{1}{1 + b_2 P_2 (1 - \theta_1)}.$$

If the effect of the second species is measured by this ratio of the binding of the labelled species, it can be seen that this decreases as P_2 is increased but depends also on the proportion of receptors occupied by the labelled species in the controls (θ_1). For a particular value of θ_1 the relation between the fraction bound, θ_1'/θ_1, and the concentration of the second species, P_2, is a hyperbola and the graphs of fraction bound against $\log P_2$ are a reverse S-shape (Figure 4.5) and the concentration reducing adsorption to 50 per cent depends greatly on the value of θ_1. If the effect of the second species is expressed as percentage inhibition of the binding of the first species, i.e. as $(\theta_1 - \theta_1')/\theta_1 \times 100$, the graph against $\log P_2$ is S-shaped and no longer reversed.

A special situation, which is common in *radioimmunoassay*, is when both species are the same but one is radio-labelled and the other is not. The binding material is a protein which has been developed immunologically and should have a high affinity for a specific substance, such as a hormone, prostaglandin or cyclic nucleotide. The concentration of this substance in a sample is estimated by the inhibition of the binding of labelled substance. If, for instance, the adsorption coefficient is known, the value of θ_1 for

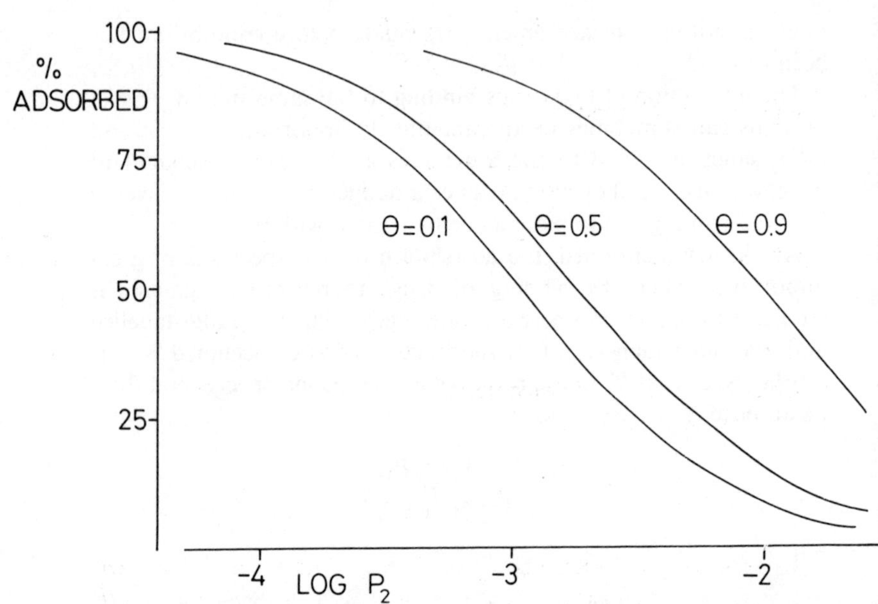

FIGURE 4.5 The effect of one species on the binding of another when both compete for the same site. The percentage of the first species adsorbed is plotted against the log concentration of the second species ($\log P_2$) for various proportions of receptor occupancy by the first species (θ_1): $b_2 = 1000$.

the concentration of (labelled) substance in the controls could be calculated and the fraction bound in the presence of the sample, θ'_1, could be used to calculate P_2. The adsorption coefficient could be calculated from measurements with a range of concentrations of labelled substance but it is more usual to measure the fraction bound using a fixed concentration of labelled substance and a range of concentrations of unlabelled substance. This standard curve can then be used for the estimation of unknown samples by a bioassay type of experiment (as in Chapter 1). Although this wastes information because it ignores the theoretical basis for the experimental curve, it may be necessary to do this because the binding protein may not be entirely specific, or it may be appreciably unstable and the standard curve may vary with time.

Instead of plotting the fraction bound, θ'_1/θ_1 against $\log P_2$ there are advantages in plotting the logit of the fraction bound, i.e. $\log \theta'_1/(\theta_1 - \theta'_1)$ against $\log P_2$. This should be a straight line because:

$$\frac{\theta'_1}{\theta_1 - \theta'_1} = \frac{1 + b_1 P_1}{b_2 P_2}$$

so logit (fraction bound)

$$= \log\left(\frac{1 + b_1 P_1}{b_2}\right) - \log P_2.$$

If you use logits from tables, note that these may be to the base e.

EXAMPLE 4:1.

Calculate values of θ for different values of P and with $b = 10^3$, $n = 1$, using the formula,

$$\theta = \frac{bP^n}{1 + bP^n}.$$

Choose values which show the shape of the curve well (ten points should suffice). Plot θ against P, θ against $\log P$, θ/P against θ, and logit (θ) against $\log P$.

Repeat the calculations with $n = 2$ and $n = 4$. Plot θ against P and logit (θ) against $\log P$. Note that the graph of θ against P becomes sigmoid but that if there are appreciable errors attached to the estimates of θ, it will be difficult to detect when $n = 2$. Note also that the graph of $\log \theta$ against $\log P$ is sigmoid when $n = 1$.

EXAMPLE 4:2.

The table shows values of the partial pressure of oxygen and the percentage saturation of haemoglobin:

Partial pressure (mm Hg)	10	15	20	25	30	35	40	50	60	100
% Saturation		27.5	41	60	69.5	75	79.5	83	85.5	91	98.5

Use a Hill plot to calculate n and K in the expression $Kp^n = y/(1-y)$. Does n appear to be constant?

(From results of Hill *Journal Physiol.* (1910), no. 40, p. vi (Proceedings).)

EXAMPLE 4:3.

The binding of $(+)$

$$\underset{C_6H_{11}}{\overset{Ph}{\diagdown}}\underset{COOCH_2CH_2\overset{+}{N}Et_3}{\overset{OH}{\diagup}}C$$

to membrane fragments from rat brain containing 'muscarinic' acetylcholine receptors was investigated by competition with the binding of (^3H)-propylbenziloylcholine

$$\underset{Ph}{\overset{Ph}{\diagdown}}\underset{COOCH_2CH_2\overset{+}{N}Me_2Pr}{\overset{OH}{\diagup}}C$$

and the following results were obtained:

0	1 nM	3 nM	10 nM	20 nM	30 nM	100 nM	300 nM	1 µM	3 µM	10 µM	Q
1623	1448	1277	1007	744	594	357	280	227	222	197	216
1587	1348	1278	917	694	646	396	282	245	238	237	228
1597	1405	1302	876	665	612	381	278	233	240	207	214
1618	1400	1285	900	691	574	390	287	227	223	230	224

The numbers indicate the counts per minute (c.p.m.) for the bound labelled ligand in the presence of a range of concentrations of the unlabelled ligand with each experiment made in quadruplicate. The temperature was 0°C. The column headed 'Q' indicates nonspecific binding to sites other than receptors, determined in the presence of $1\,\mu M$ quinuclidyl-benzilate. Calculate the average specific binding to the muscarinic receptor and the percentage inhibition of binding produced by the unlabelled ligand. The labelled ligand is present in a concentration of $0.1\,nM$ and this is known to occupy less than 1% of the receptors, calculate the affinity constant of the unlabelled ligand. What is the slope of the Hill plot?

(From results of Barlow, Birdsall and Hulme.)

EXAMPLE 4:4.

To estimate the amount of LH (luteinizing hormone) in rat pituitary glands, the binding of (^{131}I)-LH to an immunologically prepared antiserum was measured in the presence of standard solutions of LH and of homogenates of pituitary gland. The following results were obtained, expressed as counts per 100 seconds for bound ligand.

LH		Pituitary homogenate	
324 ng	64	1/300 gland	38
	64		63
	56		42
108 ng	306	1/900 gland	163
	301		191
	256		194
36 ng	650	1/2700 gland	504
	602		539
	655		434
0 ng	825		
	890		

Plot the percentage of counts bound (or the percentage inhibition of binding) against the log dose and the logit percentage against the log dose. Use the amounts causing 50% inhibition to estimate the LH content of one pituitary gland. Compare the answer with that obtained using a four-point assay made with the upper and middle two pairs of doses and again with a four-point assay made with the middle and lower doses. Comment on the results.

(From results of Dr P. S. Brown.)

Enzyme and Substrate

In most situations the adsorption process is of interest because it is associated with something happening, such as the reaction between enzyme and substrate. This is usually represented:

$$E + S \rightleftharpoons ES$$

and if the concentration of substrate is S, if that of the complex is p, and if the initial concentration of enzyme is e, the amount of free enzyme at equilibrium will be $(e-p)$. The rate of formation of the complex will be $k_{+1}(e-p)S$ and the rate of breakdown will be $k_{-1}p$. The substrate constant, K_s, which is the equilibrium constant for the dissociation of the complex, is

$$K_s = \frac{k_{-1}}{k_{+1}} = \frac{(e-p)S}{p} \quad \text{and} \quad p = \frac{eS}{K_s + S}.$$

This is strictly comparable with $\theta = bP/(1+bP)$ (page 118), because K_s is a dissociation constant and b is an association constant $(=1/K_s)$.

The breakdown of the complex ES usually involves another species, such as a hydrogen-acceptor (for a dehydrogenase) or water (for a peptidase or esterase), but this can be kept in excess so the rate of the reaction will be $k_{+2}p$, where k_{+2} is the forward rate constant for the second step. In calculating p, however, this step has so far been ignored. If it is included, the rate of breakdown of complex is $k_{-1}p + k_{+2}p$, so K_s should be replaced by another constant, the *Michaelis constant*, K_m, which is $(k_{-1}+k_{+2})/k_{+1}$ and $p = eS/(K_m + S)$. If k_{-1} is very much bigger than k_{+2}, i.e. the rate of breakdown to form products is very much slower than the rate of dissociation of unchanged substrate, there will be little difference between K_m and K_s.

The rate of the reaction is

$$k_{+2}p = \frac{k_{+2}eS}{K_m + S}$$

which has exactly the same form as the expression for adsorption (page 119),

$$x = \frac{kbP}{1+bP} : \quad \text{rate} = \frac{k_{+2}e(S/K_m)}{1+(S/K_m)}.$$

At high substrate concentrations $S/(K_m + S) \to 1$ and $p \to e$ so the reaction should proceed at its maximum rate, $V_{max} = k_{+2}e$. Accordingly the velocity,

$$V = \frac{V_{max}}{1+(K_m/S)} = \left(\frac{S}{S+K_m}\right) V_{max}$$

and $V = \frac{1}{2}V_{max}$ when $S = K_m$, i.e. K_m is the concentration of substrate at which the rate is half-maximal.

Because the rate of the reaction is directly proportional to the amount of material adsorbed, the hyperbolic relation between rate and substrate concentration can be treated in exactly the same way as the relation between amount adsorbed and partial pressure or concentration (page 119). The only differences arise from the custom of using dissociation constants with the enzyme reaction and association (affinity) constants in the adsorption process.

Transformation (i) was used by Lineweaver and Burk[6] and

involves plotting $1/V$ against $1/S$. This should give a straight line with a slope of K_m/V_{max} and when

$$\frac{1}{V} = 0, \qquad \frac{1}{S} = -\frac{1}{K_m}.$$

As has already been indicated, this method has many disadvantages. The extrapolation is most affected by the results at low concentrations of substrate, which are least accurate, and it is not easy to set statistical limits to the estimate of K_m from the variance of the results. For this kind of analysis it is desirable to choose substrate concentrations whose reciprocals increase in an arithmetic progression, so that the points are distributed evenly along the line.

Transformation (ii) was used by Augustinsson[7] and Hofstee[8] and is the same as that used by Scatchard for the binding of molecules to proteins. It involves plotting V/S against V. The equation is written in the form:

$$\left(\frac{V}{S}\right)K_m = V_{max} - V$$

so the slope is $-(1/K_m)$. This avoids extrapolation and also has the advantage that the standard error of the estimate of K_m can easily be calculated (see Appendix). It has the disadvantage that the larger experimental error, that involved in the measurement of the velocity, affects both parameters which are being plotted.

Transformation (iii) involves writing the equation in the form:

$$K_m = \left(\frac{S}{V}\right)V_{max} - S$$

and plotting S/V against S. When $S/V = 0$, $S = -K_m$ so this method has the disadvantages associated with extrapolation and with the difficulty of setting statistical limits to the estimate of K_m.

The equation can also be written:

$$\frac{V}{V_{max}} = \frac{1}{1 + (K_m/S)}$$

so

$$\text{logit}\left(\frac{V}{V_{max}}\right) = \log\left(\frac{V}{V_{max} - V}\right) = \log S - \log K_m$$

and when

$$\text{logit}\left(\frac{V}{V_{\text{max}}}\right) = 0,$$

$\log S = \log K_m$. This is the same as calculating K_m from the concentration of substrate which produces half the maximum rate but if the results are fitted by least-squares to the graph of $\text{logit}(V/V_{\text{max}})$ against $\log S$, statistical limits can be set to the estimate of $\log K_m$. The Hill plot needs an accurate estimate of V_{max}, which may be difficult if the reaction is inhibited by excess substrate, as in the hydrolysis of acetylcholine by acetylcholinesterase. There is also the disadvantage that an overestimate of V will lead to an underestimate of $(V_{\text{max}} - V)$ and consequently an even bigger overestimate of the ratio $V/(V_{\text{max}} - V)$. Similarly an underestimate of V leads to a big underestimate of the ratio. The method, however, is extremely useful for investigating whether results fit the Langmuir isotherm or if there is evidence for cooperative interactions between substrate and enzyme-substrate complex.

The linear transformation may be avoided by direct fit of the results to the hyperbola and the calculation of the substrate concentration producing half the maximum rate. It is difficult to set statistical limits to the estimate but this direct fit to the hyperbola should give a good estimate of V_{max}, and may conveniently be used in conjunction with the Hill plot.

For example, the rates of oxidation of various concentrations of tyramine, tryptamine and 5-hydroxytryptamine by amine-oxidase from guinea-pig liver are shown in Table 4.4. In this reaction two hydrogen atoms are removed from the substrate and form water by combination with molecular oxygen and the amine is converted into an imine (Schiff's base):

$$-CH_2 - NH_2 + \tfrac{1}{2}O_2 \rightarrow -CH = NH + H_2O$$

The reaction was followed by the change in the pressure of the oxygen in the system, which was converted into a change in volume, and the rate was expressed as μl oxygen absorbed in 9 minutes. The estimates of K_m and V_{max} obtained by the Lineweaver-Burk and Augustinsson methods and by direct fit to the hyperbola are compared in the lower part of Table 4.4. There is reasonable agreement between the estimates obtained by the last two methods; those obtained by the Lineweaver-Burk plot are appreciably

TABLE 4.4: *The rates of oxidation (μl oxygen absorbed in 9 minutes) of tyramine, tryptamine and 5-hydroxytryptamine by amine-oxidase from guinea-pig liver obtained with various concentrations of substrate at 37°C were:*

Concentration	Tyramine	Tryptamine	5-Hydroxytryptamine
1.0 mM	18.2	12.8	12.1
2.0	22.6	13.6	14.1
5.0	28.8	19.2	19.0
10.0	31.6	21.6	19.2

Estimates of K_m and V_{max} obtained by various methods were:

	K_m	V_{max}	K_m	V_{max}	K_m	V_{max}
Lineweaver-Burk	0.87 mM	33.6	0.76 mM	21.5	0.74 mM	20.7
Augustinsson	0.92	34.1	1.03	23.4	0.84	21.3
Direct fit to hyperbola	0.94	34.3	1.04	23.2	0.81	21.1

(From results of Barlow, *British Journal Pharmacol.* (1961), no. 16, p. 156.)

different. The values of K_m for the three compounds are similar but the rate constant for the breakdown of the complex is considerably bigger with tyramine than with the other two compounds. If a Hill plot is made of the results, the points do not lie very satisfactorily on straight lines. The graphs indicate the large errors attached to the points, however, rather than evidence for departure from the Langmuir isotherm.

The results are unsatisfactory because in none of the experiments was the concentration of substrate less than the Michaelis constant. There are experimental reasons for this failure to work with lower substrate concentrations. An oxygen uptake of $11.2 \times (310/273) = 12.7\,\mu l$ at 37°C and 1 atmosphere pressure corresponds to the reaction of $1\,\mu mol$ of substrate, which is the amount of substrate present in 1 ml of a 1 mM solution. It is difficult in these experiments to work with more than about 4 ml of solution, because of the size of the flasks used. With concentrations less than 1 mM, therefore, the substrate will have been largely used up by the time a reasonably detectable amount of oxygen has been absorbed. To follow an enzymic reaction over an adequate range of substrate concentrations more sensitive methods are desirable but there is always likely to be a conflict between the need to have the rate fast enough so that it can be measured accurately but not so fast that depletion of substrate limits the lower range of the concentrations of substrate which can be studied.

Enzyme Activity. In addition to estimating the Michaelis constants of different substrates and comparing their maximum rates, it is often necessary to be able to express enzyme activity in some way so that the properties of different preparations or tissues can be compared. The amount of enzyme which produces one mole of product in one minute is defined as 1 unit. Accordingly the value $V_{max} = 21.1\,\mu l$ oxygen in 9 min corresponds to $21.1/12.7\,\mu mol$ in $9\,min = 0.18\,\mu mol\,min^{-1}$ or 0.18 microunits. The *specific enzymic activity* is the number of units per milligram of enzyme protein. In the experiments with amine-oxidase a crude preparation of mitochondria was used and the amount of enzyme present was not known. If the molecular weight of the enzyme is known, the *catalytic constant* can be calculated, which is the moles of product formed per minute by a mole of pure enzyme. If the number of active centres in each molecule of enzyme is known, the *turnover number* can be calculated, which is the catalytic constant for each active centre. In all these calculations, however, it is essential to specify the substrate which is being used, as well as the temperature, pH and other conditions such as ionic strength.

EXAMPLE 4:5.

Check that the results shown in the upper part of Table 4.4 give the answers shown in the lower part.

Enzyme and Two Substrates

Many methods of following an enzymic reaction, such as by measuring the uptake of oxygen, the evolution of carbon dioxide, or the production of hydrogen ions, will not distinguish between the contributions from two substrates added simultaneously and only measure total effects. From the results in Table 4.4, for instance, it might be expected that if both tyramine and tryptamine were present in $1\,mM$ concentration the total rate would be $18.2 + 12.8 = 31\,\mu l$ oxygen absorbed in 9 min. The rates cannot, however, be added together in this way because the proportion of complex formed by one species will be reduced by the presence of the second species. The adsorptions are not additive unless they are very small.

The proportion of sites occupied by the first species in the presence of the second species,

$$\theta'_1 = \frac{b_1 P_1}{1 + b_1 P_1 + b_2 P_2}$$

(page 130) and if this is written in terms of substrate concentrations and substrate dissociation constants, the rate of reaction of the first species in the presence of the second species,

$$V'_1 = \left[\frac{(S_1/K_{1s})}{1 + (S_1/K_{1s}) + (S_2/K_{2s})} \right] V_{1\,\mathrm{max}}.$$

For tyramine and tryptamine there are estimates of the Michaelis constants which are 0.95 and 1.04 mM respectively and estimates of V_{max} of 34.3 and 23.2 µl/9 min, respectively (Table 4.4). If the Michaelis constant is approximately the same as the substrate constant, K_s, $S_1/K_{1s} = 1.053$ and $S_2/K_{2s} = 0.962$, so the proportion of sites occupied by the first species will be

$$\frac{1.053}{1 + 1.053 + 0.962} = \frac{1.053}{3.015} = 0.349.$$

In the absence of the second substrate the proportion is

$$\frac{1.053}{1 + 1.053} = 0.513$$

(if this proportion is calculated from the relative rates,

$$\frac{V}{V_{\mathrm{max}}} = \frac{18.2}{34.2},$$

the answer, 0.531, is slightly different because the observed value, 18.2, is not exactly the same as the least-squares calculated value, which is 17.6). The rate of reaction for the first species, V'_1, will therefore be $0.349 \times 34.3 = 12.0$ µl/9 min.

In a similar way it can be calculated that the proportion of sites occupied by the second species is $0.962/3.015 = 0.319$, compared with 0.490 in the absence of the first species. The rate of reaction for the second species, V'_2, will be $0.319 \times 23.2 = 7.4$ µl/9 min. The combined rate will therefore be 19.4 µl/9 min, which is much less than the sum of the separate rates (31 µl/9 min). It is, in fact, only slightly larger than the rate for tyramine alone (18.2 µl/9 min). If the second species had the same K_m as tryptamine but V_{max} was 15 µl/9 min, the total rate would be $12.0 + 0.319 \times 15.0 = 16.8$ µl/9 min, which is less than the rate for tyramine alone. The second species would be clearly inhibiting the reaction of the first species.

Competitive inhibitors can be regarded as a second species for which V_{max} is zero. In a series of compounds, however, there is likely to be a range of values of V_{max}, as well as of K_m, and a number of poor substrates with a small V_{max} which will inhibit the reaction of good substrates with a large V_{max}. The division between substrate and inhibitor, therefore, may not be clear-cut. Poor substrates may also be inhibitors but their effects depend also on concentration.

EXAMPLE 4:6.

Check that the effects of the second species depend on the concentrations by calculating the combined rates for tyramine and tryptamine when both are present in 2.0 mM, 5.0 mM and 10.0 mM. Use the results in Table 4.4 and the values of K_m and V_{max} already taken for calculating the combined rates at 1.0 mM. Compare the combined rate with that for tyramine alone at each concentration.

Enzyme, Substrate and Inhibitor

If the second species is not a substrate at all ($V_{max} = 0$), S_2 can be replaced by I and K_{2s} can be replaced by K_i, the dissociation constant of the enzyme-inhibitor complex, so the rate of the reaction in the presence of the inhibitor,

$$V' = \left(\frac{S_1/K_m}{1 + S_1/K_m + I/K_i} \right) V_{max}$$

if the inhibition is competitive. The graph of $1/V'$, against I, will therefore be a straight line (Figure 4.6A) with a slope of $K_m/(S_1 K_i V_{max})$. If a second set of results is obtained with a different substrate concentration, S_2, a second straight line will be obtained which intersects the first where

$$V' = \left(\frac{S_1/K_m}{1 + S_1/K_m + I/K_i} \right) V_{max} = \left(\frac{S_2/K_m}{1 + S_2/K_m + I/K_i} \right) V_{max}$$

Accordingly

$$\frac{1}{S_1}(1 + I/K_i) = \frac{1}{S_2}(1 + I/K_i)$$

and because S_1 does not equal S_2, $1 + I/K_i = 0$, so the value of I corresponding to the point of intersection $= -K_i$.

If the process is noncompetitive and such that

$$\theta = \frac{b_1 P_1}{1 + b_1 P_1}\left(\frac{1}{1 + b_2 P_2}\right) \quad \text{(page 131)}$$

the velocity

$$V' = V_{max}\left(\frac{S/K_m}{(1 + S/K_m)(1 + I/K_i)}\right)$$

and the graph of $1/V'$ against I will again be a straight line. When $1/V' = 0$, $(1 + I/K_1)$ must also be zero, because $[1 + (S/K_m)]$ cannot be zero, so $I = -K_i$. Lines with all other concentrations of substrate must intersect at this point (Figure 4.6B).

In many experiments the concentration of inhibitor which reduces the reaction rate to 50% is measured and is called I_{50}. For a competitive inhibitor the rate of the inhibited reaction, V', is related to that of the uninhibited reaction, V', by the equation

$$\frac{V'}{V} = \frac{1 + S/K_m}{1 + S/K_m + I/K_i},$$

so the graph of V/V' against I should be a straight line (Figure 4.6C) and for 50% inhibition $I_{50}/K_i = 1 + S/K_m$ and $I_{50} = K_i(1 + S/K_m)$. Accordingly I_{50} is not a constant but depends on S/K_m, i.e. on the nature of the substrate (which determines K_m), and on the concentration. It will increase if the substrate concentration is increased.

For a noncompetitive inhibitor

$$V'/V = \frac{1}{1 + I/K_i}$$

so for 50% inhibition $I_{50} = K_i$, i.e. it is independent of S.

For a competitive inhibitor

$$V/V' = 1 + \frac{I/K_i}{1 + S/K_m} \quad \text{so} \quad \frac{V - V'}{V'} = \frac{IK_m}{K_i(S + K_m)}$$

or

$$\frac{IV'}{V - V'} = K_i(1 + S/K_m).$$

The graph of $IV'/(V - V')$ against S should be a straight line (Figure 4.6D) with a slope of K_i/K_m and when

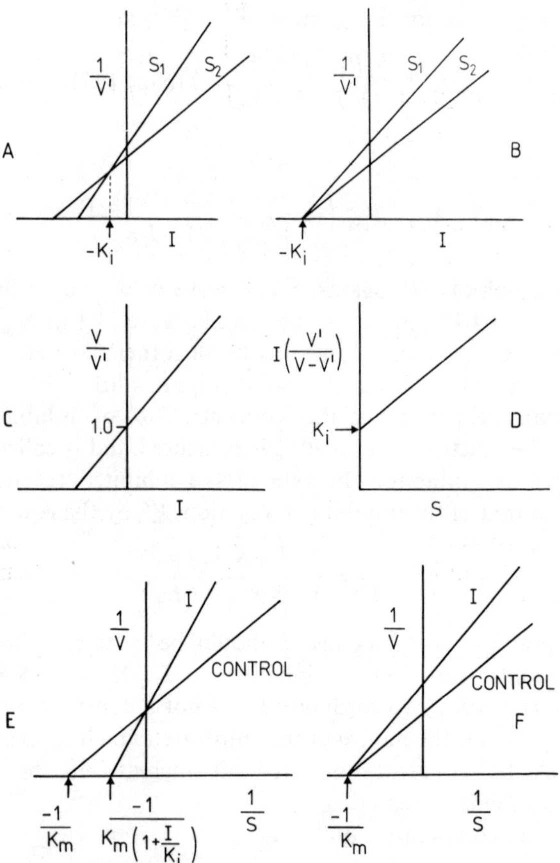

FIGURE 4.6 Effects of inhibitors; V' is the velocity of the inhibited reaction.

(A) Competitive inhibition:

$$\frac{1}{V'} = \frac{K_m}{SV_{max}}\left(1 + \frac{1}{K_i}\right) + \frac{1}{V_{max}}.$$

Note that extrapolation is necessary to calculate K_i (also that $S_2 > S_1$).

(B) Noncompetitive inhibition:

$$\frac{1}{V'} = \frac{1}{V_{max}}\left(1 + \frac{K_m}{S}\right) + \frac{I}{K_i V_{max}}\left(1 + \frac{K_m}{S}\right).$$

Note that if any inhibitor is present at all ($I > 0$) V' can never equal V_{max}, however large the value of S (also that $S_2 > S_1$).

(C) Competitive inhibition: relative velocities of control (V) and inhibited (V') reactions:

$$\frac{V}{V'} = 1 + \frac{IK_m}{K_i(S+K_m)}.$$

$$S = 0, \qquad \frac{IV'}{V - V'} = K_{\mathrm{i}}.$$

For a noncompetitive inhibitor $V/V' = 1 + I/K_{\mathrm{i}}$, so

$$\frac{V - V'}{V'} = \frac{I}{K_{\mathrm{i}}} \quad \text{and} \quad \frac{IV'}{V - V'} = K_{\mathrm{i}}$$

and is constant and independent of S.

For a competitive inhibitor a Lineweaver-Burk plot of $1/V'$ against $1/S$ will again be a straight line because

$$\frac{V_{\max}}{V'} = \frac{K_{\mathrm{m}}}{S} + 1 + \frac{IK_{\mathrm{m}}}{SK_{\mathrm{i}}} = 1 + \frac{K_{\mathrm{m}}}{S}\left(1 + \frac{I}{K_{\mathrm{i}}}\right)$$

but when

$$\frac{1}{V'} = 0, \qquad S = -K_{\mathrm{m}}\left(1 + \frac{I}{K_{\mathrm{i}}}\right)$$

so the 'apparent K_{m}' obtained from the results in the presence of the inhibitor will be bigger than the true K_{m} by an amount $IK_{\mathrm{m}}/K_{\mathrm{i}}$ (Figure 4.6E).

When

$$\frac{V}{V'} = 0, \quad I = \frac{K_{\mathrm{i}}}{K_{\mathrm{m}}}(S + K_{\mathrm{m}}).$$

(D) Competitive inhibition:

$$I\left(\frac{V'}{V - V'}\right) = K_{\mathrm{i}}\left(1 + \frac{S}{K}\right).$$

When

$$S = 0, \quad I\left(\frac{V'}{V - V'}\right) = K_{\mathrm{i}}.$$

For noncompetitive inhibition $I[V'/(V - V')]$ is independent of S (i.e. constant).

(E) Competitive inhibition:

$$\frac{1}{V'} = \frac{1}{V_{\max}}\left[1 + \frac{K_{\mathrm{m}}}{S}\left(1 + \frac{I}{K_{\mathrm{i}}}\right)\right].$$

The Lineweaver-Burk plot for the results in the presence of the inhibitor gives when extrapolated the reciprocal of $K_{\mathrm{m}}[1 + (I/K_{\mathrm{i}})]$, compared with the reciprocal of K_{m} for the controls.

(F) Noncompetitive inhibition:

$$\frac{1}{V'} = \frac{1}{V_{\max}}\left(1 + \frac{K_{\mathrm{m}}}{S}\right)\left(1 + \frac{I}{K_{\mathrm{i}}}\right).$$

The Lineweaver-Burk plot gives the same value for K_{m} as the controls.

For a noncompetitive inhibitor,

$$\frac{V_{\max}}{V'} = \left(1 + \frac{K_m}{S}\right)\left(1 + \frac{I}{K_i}\right),$$

so when $1/V'$ is plotted against $1/S$, the 'apparent K_m' will not be altered by the presence of the inhibitor, because when $1/V' = 0$, $1 + (K_m/S) = 0$ so $S = -K_m$ (Figure 4.6F). The effect of such an inhibitor on the slope of the graph of $1/V$ against $1/S$, however, is the same as that of a competitive inhibitor.

If results of V' and S are fitted directly to the hyperbola

$$V' = \left(\frac{S}{S + K_m}\right) V_{\max},$$

the 'apparent K_m' and $V_{\max}$ are obtained, and for a competitive inhibitor there should be no effect on $V_{\max}$ but the 'apparent K_m' will be bigger than the K_m for the uninhibited reaction and

$$\frac{\text{apparent } K_m}{\text{control } K_m} = 1 + \frac{I}{K_i}$$

For a noncompetitive inhibitor there should be no effect on K_m but $V_{\max}$ will be reduced and

$$\frac{\text{control } V_{\max}}{\text{inhibited } V_{\max}} = 1 + \frac{I}{K_i}.$$

There are therefore a number of ways in which the inhibitor constant, K_i, can be measured but whatever method is used it is

TABLE 4.5: *Effects of n-pentyltrimethylammonium iodide on the hydrolysis of acetylthiocholine iodide by acetylcholinesterase from electric eel. The rate of hydrolysis is calculated from the change in optical density (at 412 nm) in 15 cycles ($11\frac{1}{2}$ min) multiplied by 100. Temperature 25°C; pH = 8.1.*

Acetylthiocholine concentration, mM	0.1	0.2	0.3	0.4	0.6	0.8	1.2	1.6
Controls	21.25	27.0	29.5	32.0				
	21.0	26.5	28.75	30.5				
1.25 mM *n*-pentyltrimethylammonium	6.0	12.5	15.0	17.5				
		10.75			15.75	19.75	21.5	
2.5 mM *n*-pentyltrimethylammonium		6.0			10.0	12.5	14.0	
					9.5	13.5	16.0	18.25

(From results of Barlow, *Journal Pharm. Pharmacol.* (1978), no. 30, p. 703.)

important to study a range of concentrations of inhibitor and a range of concentrations of substrate to investigate the type of inhibition as well as to estimate K_i. For example, the effects of *n*-pentyltrimethylammonium iodide on the hydrolysis of acetylthiocholine by acetylcholinesterase are shown in Table 4.5. The reaction was followed spectrophotometrically by the formation of a light-absorbing product from thiocholine, produced by enzymatic hydrolysis of the substrate, and the amount of substrate used produces a negligible effect on the concentration even with substrate concentrations as low as 0.1 mM. The least-squares fit to the hyperbola gives an estimate of $K_m = 0.0748$ mM, $V_{max} = 36.78$ for the controls, $K_m = 0.345$ mM, $V_{max} = 31.08$ in the presence of 1.24 mM *n*-pentyltrimethylammonium and $K_m = 0.629$ mM, $V_{max} = 24.92$ in the presence of 2.5 mM *n*-pentyltrimethylammonium. The inhibition is therefore not strictly competitive, because the compound reduces V_{max} but the two estimates of K_i are

$$1.25 \left/ \left(\frac{0.345}{0.0748} - 1 \right) \right. = 0.346 \text{ mM}$$

and

$$2.5 \left/ \left(\frac{0.629}{0.0748} - 1 \right) \right. = 0.337 \text{ mM}$$

which are in reasonable agreement.

The results illustrate some of the practical difficulties. If the rates are slow the sensitivity is reduced and with this spectroscopic method there is a limit to the extent to which the rates in the presence of the inhibitor may be increased by increasing the substrate concentration. The acetylthiocholine solutions always contain traces of thiocholine formed from chemical hydrolysis and with high concentrations the absorption from this makes it impossible to follow the enzymic reaction.

Although the antagonism between *n*-pentyltrimethylammonium and acetylthiocholine does not appear to be strictly competitive, it is remarkable that the slopes of the Hill plots of the results are close to unity.

EXAMPLE 4:7.

Make a Hill plot of the results in Table 4.5 using the values of V_{max} given above and calculate the slopes.

The Need for Other Models

In the models of adsorption which have so far been considered the binding site, R, is regarded as remaining unaltered during the process, as it might do if it were a solid catalytic material such as charcoal or platinum. It is possible, however, that the adsorption of a molecule will affect the properties of the site and consequently the adsorption of other molecules. This is particularly likely if the catalytic material is a protein in solution in water or even merely in contact with water. Binding at one site, for instance, might lead to changes in structure which prevent binding at another site, thus replacing the competitive model for antagonism by binding at the same site.

Sometimes the behaviour of compounds makes it clear that a single site model, R, is inadequate (e.g. if there is desensitisation, see page 173). Once the single site model has been abandoned, however, there are many possibilities and it is necessary to decide which the results fit best and indeed whether the fit is significantly better than with a single site model. With the large errors attached to most experiments it is usually extremely difficult to justify abandoning the single site model simply from the fit of the results. Sometimes, however, the graph of response (e.g. rate of reaction, amount bound) against dose is convincingly sigmoid at low concentrations, indicating departure from the Langmuir isotherm (see Example 4.2). Usually results obtained with such low concentrations are too inaccurate for this to be clear. It should be noted that it is the graph of response against dose which must be examined; the graph of response against log dose is sigmoid anyway for results which fit the Langmuir isotherm.

If the adsorption site is altering there should be changes in the affinity for the ligand and this may be detectable in a Hill plot. If there is *positive cooperativity* and the affinity is increased by the presence of molecules already bound, the intercept on the X-axis will increase and the line will become steeper so the slope, n, will be greater than 1. If there is *negative cooperativity* and the affinity is decreased by the presence of molecules already bound, the slope will be less than 1. Changes in K, in fact, explain why the slope can be fractionally variable, which is difficult to interpret if the slope is thought to depend on the number of molecules of ligand reacting with each site and K is constant (page 120).

On the single site model the Hill plot of the results should have a slope of 1 and it is remarkable how often this is true, even allowing for difficulties in estimating the maximum effect and the effects of errors on the ratio $V/(V_{max} - V)$ (page 140). A slope significantly different from 1 is not compatible with this model, so a Hill plot is usually the first step in making any decision about replacing it by a more complicated mechanism (some of which are considered later in this chapter).

Agonists and Receptors

The simplest hypothesis for the adsorption of a drug at a receptor is:

$$A + R \rightleftharpoons AR$$

so the process may also be expected to fit the Langmuir isotherm. If the drug is present in a concentration, A, and occupies a proportion, y, of the receptors and K is the *affinity* (association) *constant*, application of the Law of Mass Action gives:

$$AK = \frac{y}{1-y} \quad \text{or} \quad y = \frac{AK}{1 + AK}.$$

It is not known, however, how receptor occupancy, y, is related to the size of the biological response which is actually measured.

If the response, R, is directly proportional to y, $R = \alpha y$, where α is a constant, just as $V = k_{+2}p$ for an enzyme. The maximum response would be obtained when $y = 1$ and the concentration of drug producing half the maximum response, EC_{50}, would be $1/K$. This was considered as a possibility by Clark,[9] though he wrote '...the application of these formulae to biological data involves certain assumptions which are unproven. In the first place the formulae assume that the receptors in a cell resemble the surface of a polished metal, in that they are all equally accessible to the drug. In the second place the interpretation assumes that the amount of biological effect produced is directly proportional to the number of specific receptors occupied by the drug. Both these assumptions seem improbable...' Nevertheless, some workers have supposed that the affinity of an agonist can be measured from EC_{50} and the term 'pD_2' has been applied to $-\log EC_{50}$.[10]

Although each system producing a biological response must be considered separately, it is clear that with some of them the size of the response is *not* directly proportional to y.

With many systems this can be seen from the slope of the log

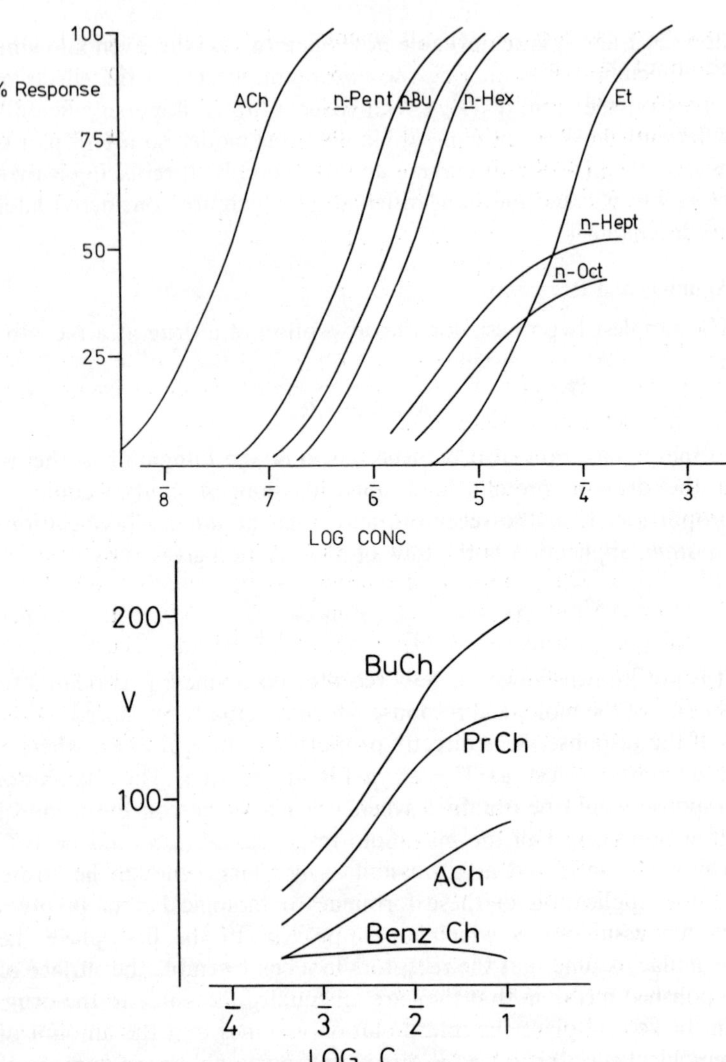

FIGURE 4.7 A comparison between *A*. the graphs of response (shortening of a piece of guinea-pig ileum) against log concentration of agonist and *B*. the rates of hydrolysis of esters of choline by butyrylcholinesterase against log concentration of substrate.

In *A* idealised results are shown for acetylcholine (ACh) and *n*-alkyltrimethyl-ammonium salts (RNMe$_3$), based on the work of Stephenson (1956) *Br. J. Pharmacol.*, no. 11, p. 390. Note that five compounds produce the same maximum response even though they differ 1000-fold in the concentrations required to produce comparable effects. A class experiment for obtaining similar results is described in 'Pharmacological Experiments on Isolated Preparations', *Edinburgh Staff*, 1972, p.

dose response curve. As already mentioned (Chapter 1, page 15) the graph of response against log dose is usually S-shaped and is roughly linear in the middle range. If the response is directly proportional to y, some idea of the slope of this part could be obtained by calculating the ratio of the concentrations producing 10% and 90% receptor occupancy ($A_{10\%}$ and $A_{90\%}$). Because $AK = y/(1-y)$,

$$\frac{A_{90\%}}{A_{10\%}} = \left(\frac{0.9}{0.1}\right)\Bigg/\left(\frac{0.1}{0.9}\right) = \left(\frac{0.9}{0.1}\right)^2 = 81.$$

For 25% and 75% occupancy the ratio is 9. In practice, however, there are many systems where the graph of response against log dose is much steeper than this.

With many types of recording it is likely that there is some mechanical limit which determines the maximum response. For example, with substances which slow the heart, such as acetylcholine or carbachol, quite a small increase in concentration from that which produces detectable slowing will stop the heart altogether. Similarly with the contraction of smooth muscle recorded isotonically (change in length with constant load) there is likely to be a mechanical limit to the shortening which is possible, so the log dose response curve is steep. The use of isometric recording (change in tension with constant length) often produces a considerably flatter curve with the same tissue, so the maximum response is more likely to be receptor-limited, rather than tissue-limited, but this is by no means certain.

A particularly important set of results has been obtained by Stephenson[11] with acetylcholine ($CH_3COOCH_2CH_2\overset{+}{N}Me_3$) and alkyltrimethylammonium salts ($CH_3(CH_2)_n\overset{+}{N}Me_3$) tested on the ileum of the guinea-pig, set up as an isolated preparation with the contractions recorded isotonically. Many of the compounds give the same maximum response, though with quite different concentrations. It is only when the chain length is increased from n-hexyl to n-heptyl that the maximum response obtainable becomes

---

64. Usually the partial agonists produce smaller effects than those shown here, sometimes nothing at all, but this varies from preparation to preparation.

In *B* the substrates are butyrylcholine (BuCh), propionylcholine (PrCh), acetylcholine (ACh) and benzoylcholine (BenzCh); note that they all have different maximum velocities.

(These results are taken from the work of Augustinsson, *Acta Physiol. Scand.* (1948), no. 15, Supplement 52.)

obviously smaller than that produced by the shorter alkyltrimethyl-ammonium compounds and acetylcholine (Figure 4.7). The results could be taken to indicate that increasing the length of the alkyl chain beyond n-pentyl lowers the affinity but that it is only with the n-heptyl compound that the proportionality constant, α, decreases. A decrease in affinity with chain length would be surprising, however, because it is found that with longer chains the compounds are antagonists and their affinity increases with chain length. It is also remarkable that the proportionality constant should apparently be the same for so many compounds, in contrast to what is found with substrates in enzymic reactions, where $V_{\max}$ varies greatly from one substrate to another.

These results suggest that above a certain point the maximum response is determined by the tissue. The maximum response obtainable with less active drugs may be less than this limit set by the tissue and substances such as n-heptyltrimethylammonium behave as *partial agonists* on this preparation, producing some effect but never the maximum of which the tissue is capable. With the more active drugs which are *full agonists*, it is possible that this maximum response of the tissue is produced when the proportion of receptors occupied is quite small; for acetylcholine and the guinea-pig ileum it may be only 1 %. In these conditions,

$$AK = \frac{y}{1-y} \rightarrow AK = y,$$

and the relationship between y and response will therefore have the same form as the experimentally observed relationship between dose and response (Figure 4.8). This is not a straight line, such as would be expected for direct proportionality, but a curve which flattens at higher responses. This indicates that a particular increase in y produces much smaller effects at high responses than at low ones. Conversely, to produce a particular increase in response you need to activate more receptors when the effect is already large than when it is small.

More evidence that active agonists produce a maximum response when only a small proportion of receptors is occupied is obtained from experiments with irreversible antagonists, which are discussed later (page 162).

To account for differences between agonists Stephenson[11] suggested that the response, R, is a function of the *stimulus*, S, where $S = ey$ and e, the *efficacy*, is a measure of the ability of the

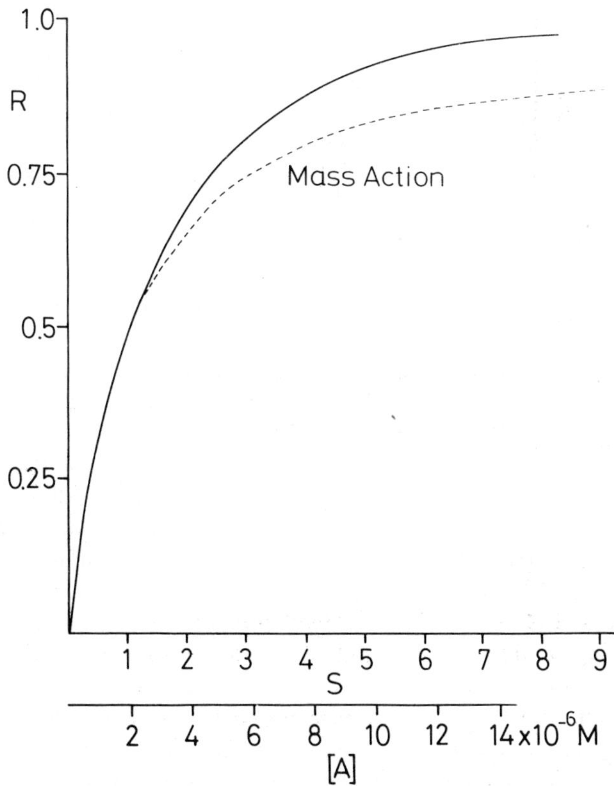

FIGURE 4.8 The response, R, expressed as a fraction of the maximum obtainable, plotted against the concentration of n-butyltrimethylammonium, $[A]$, and against the stimulus, S, where 1 unit of stimulus produces a half-maximum response. The assumption is made that only a small proportion of receptors is occupied, so the receptor occupancy, y, is directly proportional to the concentration, A.

The ratio $(A_{75})/(A_{25})$ is about 7.5. The broken line indicates the mass action relationship between receptor occupancy, y, and relative concentration.

drug to activate the receptors. He defined 1 unit of stimulus as that needed to produce a half-maximum response and if the agonist only occupies a small proportion of receptors, y is directly proportional to the concentration of agonist, A, and the relation between response, R, and stimulus, S, is obtained directly from the experimental curve (Figure 4.8). This relationship is highly individual; it refers to the particular piece of tissue and the particular conditions of the experiment. It may change with time, for instance, as sensitivity to the drugs increases of decreases. The system is extremely complex, involving the contraction of the muscle as well as the interaction between the drug and the receptor, and it is not possible to produce any satisfactory theoretical basis for the shape of this curve. It is, however, clear that the situation is different from what is observed with simple enzymes; the response is *not* directly proportional to receptor occupancy.

With a particular concentration of the substrate of an enzyme the rate of the reaction depends on two factors, K_m and V_{max}, which it is usually fairly easy to estimate. With a drug producing an effect by acting on receptors in a tissue, the response produced by a particular concentration also depends on two factors, affinity and efficacy, but these are not easy to estimate and there is the added complication that the relation between response and stimulus may vary from one preparation to another. The relative activities of compounds can be assessed from the ratios of the concentrations which produce comparable responses but it is not possible to say whether the more active compounds have higher affinity for receptors or higher efficacy (activate the receptors better when absorbed). Methods for attempting to measure the affinity of an agonist are discussed later (page 169).

It is remarkable that with many agonists acting at the same receptors, the effects are additive. If the effects produced by a concentration, A_1, of one agonist match those produced by A_2 of a second, $A_1 + A_2$ together produce the same effect as $2A_1$ or $2A_2$ (the log dose response curves are parallel). This is different from what is usually found with two substrates of an enzyme; the substrate-rate curves are not likely to be parallel and the effects are not likely to be strictly additive (page 142). If one of the compounds, however, is a partial agonist with a low efficacy and requires a large receptor occupancy to produce its effects, it will not act additively with a full agonist and the responses may be used, in fact, to calculate the affinity of the partial agonist for the receptors (see below). The additive nature of agonist effects is common, but not a general rule. It is, however, assumed in a number of situations, such as in the calculation of the activity of mixtures of isomers (page 88). If possible, it is important to check whether it is justified. If there is a big difference in the activity of the isomers, one of them may well be a partial agonist.

Antagonists and Receptors

Although the relation between the biological response and the adsorption of agonists is complex, that between the biological response and the adsorption of antagonists is much simpler. The simplest model for the antagonism between the agonist A and the antagonist B is the same as that used for competitive inhibition of an enzyme:

$$A + R \rightleftharpoons AR \rightarrow \text{response}$$
$$B + R \rightleftharpoons BR$$

If the agonist is present in a concentration, A, and occupies a proportion, y, of receptors and has an affinity constant, K_a, and the antagonist is present in a concentration, B, occupies a proportion, z, of receptors and has an affinity constant, K_b, the proportion of unoccupied receptors will be $(1 - y - z)$ and

$$AK_a = \frac{y}{1-y-z}, \quad BK_b = \frac{z}{1-y-z} \quad \text{and} \quad \frac{AK_a}{BK_b} = \frac{y}{z}.$$

But

$$z = (1 - y - z)BK_b = \frac{(1-y)BK_b}{(1+BK_b)},$$

so

$$AK_a = \frac{y}{1-y}(1 + BK_b) \qquad \text{(Gaddum[12]).}$$

This is strictly comparable with the equation

$$b_1 P_1 = \frac{\theta_1}{1-\theta_1}(1 + b_2 P_2) \qquad \text{(page 130).}$$

It indicates the effects of the antagonist on the occupancy of receptors by agonist but in itself it does not show how the size, R, of the response would be reduced, because the relationship between R and y is not understood. The effect of an antagonist on the size of the response, e.g. expressed as the percentage reduction in the response, is, in fact, unpredictable, because it depends on the slope of the graph of response against log dose. The percentage reduction will therefore vary from one preparation to another and is a most unsatisfactory way of attempting to measure antagonism.

If the concentration of agonist is increased, however, the effect of the antagonist should be reversed. If a response, R, was obtained by a concentration, a, of agonist alone, this concentration will produce a much smaller response (or none at all) in the presence of a concentration, B, of antagonist, but the response may be restored to its original size by increasing a to some stronger concentration, A. If the responses are the same, the proportion of receptors occupied, y, should be the same because it is the same agonist which is producing the responses. When the antagonist is present, $AK_a = [y/(1-y)](1+BK_b)$ and when the agonist is alone, $aK_a = y/(1-y)$, so $A/a = 1+BK_b$. The effect of a particular concentration of an antagonist may therefore be expressed as this ratio, A/a, which is called the *dose-ratio, DR*, and the relation $DR = 1+BK_b$ is known as the Gaddum-Schild equation.[13]

Schild has used the symbol pA_x for the logarithm of the reciprocal of the concentration of antagonist producing a dose-ratio x; the number will become smaller as x is bigger, because it is necessary to use more antagonist in order to produce a bigger dose-ratio. When the dose-ratio is 2, $B = 1/K_b$ so pA_2 is the same as $\log K_b$.

The symbol, pD_2, is sometimes used in a similar way, as if it were equivalent to $\log K$ for an agonist (page 151). This will only be true if the response is directly proportional to receptor occupancy and is maximum when all receptors are occupied, which is usually unlikely and often demonstrably wrong. Values of pD_2 are not likely to be an estimate of affinity, therefore, but they are useful experimentally as an indication of the concentration likely to be active.

The graph of dose-ratio against antagonist concentration should be a straight line with a slope of K_b but if a wide range of concentrations has been used it is more convenient to plot $\log(DR-1)$ against $\log B$.[14] This should be a straight line with a slope of unity and when $\log(DR-1) = 0$, $\log B = -\log K_b$.

The dose-ratio produced by a particular concentration of antagonist should be independent of the size of the response, so the graph of log dose against response, which is usually sigmoid, is merely displaced towards higher concentrations by a constant amount, $\log DR$ (Figure 4.9). It should also be independent of the nature of the agonist; the same dose-ratio should be obtained with any agonist which acts at the receptors R.

The proportion of receptors occupied by antagonist, z, is likely to be much larger than that occupied by an active agonist, so

$$BK_b \doteqdot \frac{z}{1-z} \quad \text{and} \quad z = \frac{BK_b}{1+BK_b} = \frac{DR-1}{DR}.$$

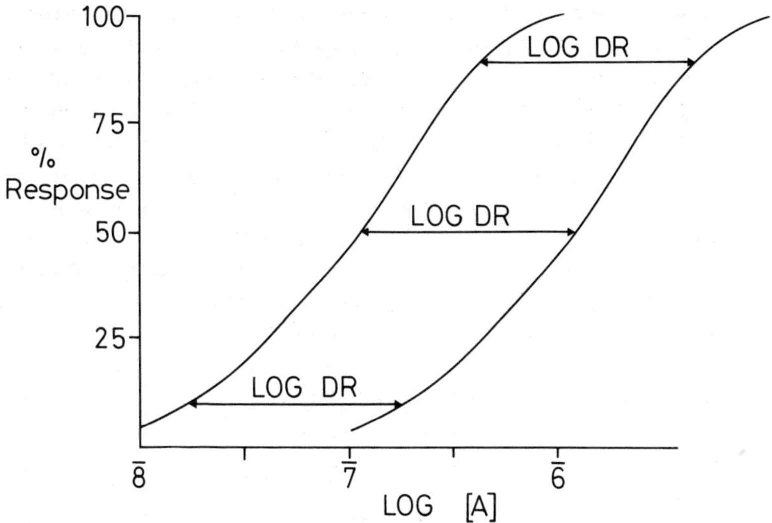

FIGURE 4.9 Effects of a competitive antagonist on the graph of response against log, concentration of agonist (log A). The increase in log concentration is 1 unit so the dose-ratio (DR) is 10. These results might well have been obtained with acetylcholine on the guinea-pig ileum alone and in the presence of 10^{-8}M atropine. Compare this picture with Figure 4.5, which shows the percentage of one species adsorbed in the presence of a second species. The adsorption process is the same in both (but the x-axis is the antagonist in Figure 4.5 and the agonist in Figure 4.9) but Figure 4.9 also includes the conversion of receptor occupancy into response, i.e. the relation shown in Figure 4.8.

For a dose-ratio of 2, therefore, about half the receptors will be occupied by antagonist and this is a much bigger proportion than that occupied by an active agonist, which might be 1 or 2% (page 162).

If a second competitive antagonist is also present

$$AK_a = \frac{y}{1-y-z_1-z_2}, \quad B_1K_{b1} = \frac{z_1}{1-y-z_1-z_2},$$

$$B_2K_{b2} = \frac{z_2}{1-y-z_1-z_2}, \quad \frac{y}{AK_a} = \frac{z_1}{B_1K_{b1}} = \frac{z_2}{B_2K_{b2}}$$

so

$$AK_a = \frac{y}{1-y-(y/AK_a)(B_1K_{b1}+B_2K_{b2})}$$

$$= \frac{y}{1-y}(1+B_1K_{b1}+B_2K_{b2}).$$

If DR_1 is the dose-ratio produced by B_1 alone and DR_2 is that produced by B_2 alone, the combined dose-ratio, DR_{12}, will be $(1+B_1K_{b1}+B_2K_{b2})$ which is DR_1+DR_2-1 (because $DR_1 = 1+B_1K_{b1}$ and $DR_2 = 1+B_2K_{b2}$). This is important in considering the effects of stereochemical impurities in partially resolved forms of compounds which are antagonists. For a mixture containing a concentration, B, of antagonist of which a proportion, y_s, is the stronger isomer, the dose-ratio observed $DR = 1+(1-y_s)BK_w+y_s BK_s$, where K_w and K_s are the affinity constants of the weaker and stronger forms, respectively. The apparent affinity constant

$$K^* = \frac{DR-1}{B} = (1-y_s)K_w+y_sK_s,$$

so even though the two forms compete with each other as well as with the agonist and the dose-ratios are not additive, the apparent affinity constant is nevertheless the sum of the contributions from the two forms. The biological activity of antagonists is additive if it is expressed in terms of affinity constants.

The combination of two antagonists can also be made use of as a test for competitive behaviour. If one of the antagonists is known to be competitive and is used in a concentration which by itself produces a dose-ratio of about 100 and the other compound is used in a dose-ratio which by itself produces a dose-ratio of 10, the combined dose-ratio should be 109. If the second antagonist is not competitive it should act independently and produce a further 10-fold antagonism on top of the 100-fold antagonism, i.e. the combined dose-ratio should be 1000.

Noncompetitive Antagonism

If the antagonism is not competitive and similar to that already considered (page 131), the proportion of receptors occupied by the antagonist is independent of the presence of the agonist. So although for the agonist

$$AK_a = \frac{y}{1-y-z},$$

$$BK_b = \frac{z}{1-z} \left(\text{not} \frac{z}{1-y-z} \right).$$

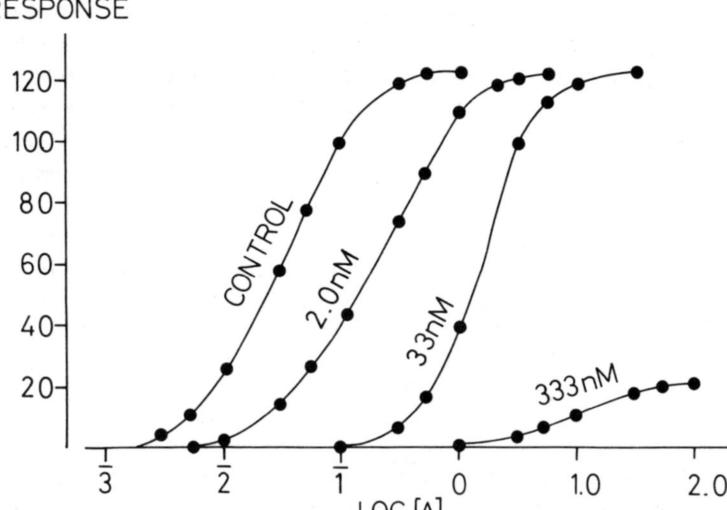

FIGURE 4.10 Antagonism of histamine by the alkylating agent, SY-28 (GD 121). The responses to histamine in the presence of 2×10^{-9} and 3.3×10^{-8}M antagonist are reasonably consistent with competition; dose-ratios of over 100 could be obtained. With 3.3×10^{-7}M antagonist the responses are clearly affected noncompetitively.

(Redrawn from results of Nickerson, *Nature* (1956), no. 178, p. 696.)

So

$$z = \frac{BK_b}{1 + BK_b}$$

and

$$y = \frac{AK_a(1-z)}{1+AK_a} = \frac{AK_a}{1+AK_a}\left(\frac{1}{a+BK_b}\right).$$

When $B = 0$, the expression becomes the same as that for the agonist alone but when $B > 0$, y can never become 1, however big the value of A. If a maximum response is produced by an agonist only when $y = 1$, an antagonist of this type would be expected to depress it irreversibly, just as in the same situation with an enzyme (page 145).

For the actions of drugs such as histamine and acetylcholine on the isolated guinea-pig ileum, however, this is not what is found. Nickerson[15] observed that the antagonism of histamine by an alkylating agent (SY-28) was apparently reversible over a very considerable range and then becomes irreversible (Figure 4.10). It is possible that the nature of the antagonism changes from a competitive action to a noncompetitive action but a more likely explanation is that histamine can produce a maximum response from the tissue when only a small proportion of receptors is occupied. If the response produced by a of agonist by itself can be produced by A of agonist in a concentration of antagonist which occupies a proportion z of the receptors, $aK_a = y/(1-y)$ and $AK_a = y/(1-y-z)$ and if z is large $(\rightarrow 1)$ there will be values of y which cannot be achieved. If y is small, however, the expression becomes $1/(1-z)$ and there will be a parallel shift of the log dose response curve, as with competitive antagonism, until y becomes comparable with $1-z$. As $z = BK_b/(1+BK_b)$, $1-z = 1/(1+BK_b)$ and the dose-ratio $DR = 1+BK_b$ so the antagonism is indistinguishable from competition at this stage. The proportion of receptors can again be approximately estimated from $(DR-1)/DR$, though this really gives the value of $z/(1-y)$. From the results it is possible to set a limit to the proportion of receptors, y, needed by an agonist to produce a maximal response. If a maximum response is still produced in the presence of a concentration of antagonist which produces a dose-ratio of 100, the proportion of receptors occupied by antagonist is 99% so the agonist can produce a maximum response with less than 1% of the receptors available.

In view of the results obtained with this type of compound it is clear that compliance with the Gaddum-Schild equation $DR = 1+BK_b$ is no proof that an antagonist is behaving competitively. The nature of its antagonism, however, can be determined by testing it together with a known competitive antagonist.

Antagonism not Involving Receptors

In many tissues there is the possibility of active physiological antagonism (page 31) because, for example, they are innervated by both sympathetic and parasympathetic nerves which produce opposite effects when stimulated. The transmitters released, noradrenaline (norepinephrine) and acetylcholine, are physiological antagonists but it will not be possible to calculate a relation between the concentrations of acetylcholine and noradrenaline for which the response is constant. The dose-ratio will depend upon the sensitivity of the particular tissue to the two compounds, i.e. the log dose response curves for the two agonists and these are probably not related to each other.

Antagonism can also occur when a drug reacts with an agonist in some way and produces a compound which is inactive. The antidotes to some poisons work in this way. For example, a compound containing thiol groups such as dimercaprol ($HOCH_2$—CHSH—CH_2SH) will react with arsenical compounds and so reduce their toxicity. If the total concentration of the agonist is A, if the concentration which is free is A_f, if the total concentration of antagonist is B and if the compounds react together so that $A + B = AB$, the concentration of the complex AB is $A - A_f$, and the concentration of free B is $B - (A - A_f)$, so the equilibrium (affinity) constant for the complex,

$$K = \frac{(A - A_f)}{A_f[B - (A - A_f)]} \quad \text{and} \quad A = A_f + \left(\frac{KA_f}{1 + KA_f}\right)B.$$

If A_f is the amount needed to produce a particular biological response, the dose-ratio

$$\frac{A}{A_f} = 1 + \left(\frac{K}{1 + KA_f}\right)B$$

so the graph of dose-ratio against the concentration of antagonist will be a straight line with a slope of $K/(1 + KA_f)$ and the results will be indistinguishable from competitive antagonism.[16] The different nature of the antagonism will be apparent if experiments are made with a second antagonist which is known to be competitive.

Antagonism by reaction is extremely important in the use of chelating agents to treat poisoning by various toxic ions but this may not involve simply one molecule of each species, as has been considered here. If the ratio is not 1:1 it should be easier to distinguish reaction from competition, though this can always be detected chemically.

EXAMPLE 4:8.

At a concentration of 10^{-8} M a competitive antagonist produces a dose-ratio of 35; calculate the affinity constant. What dose-ratios would you expect with concentrations of 10^{-9} M and 10^{-7} M?

EXAMPLE 4:9.

Estimates of $\log K$ for acetylcholine receptors of the guinea-pig ileum obtained with resolved samples of $(+)$- and $(-)$-

benzhexol were 5.700 and 8.700 respectively. What is the minimum stereochemical purity? Solutions were made up and mixed so as to obtain a known proportion of the stronger enantiomer. The concentrations tested produced the following dose-ratios:

% Stronger	Concentration	Dose-ratio
0.5	10^{-5} M	25.0
1.0	5×10^{-6}	24.9
4.8	10^{-6}	16.5
9.1	5.5×10^{-7}	24.6
16.7	2.5×10^{-7}	22.4
50.0	1.2×10^{-7}	28.1

Calculate the dose-ratios which the enantiomers should produce alone and the combined dose-ratios if they act competitively and if they act non-competitively. Calculate the apparent affinity constant, K^*, for each concentration and plot $\log(K^* - K_w)$ against $\log(y_s)$ where K_w is the affinity constant of the weaker isomer and y_s is the percentage of the stronger isomer present. Are the compounds behaving competitively?

(From results of Barlow, Franks and Pearson, *Journal Pharm. Pharmacol.* (1972), no. 24, p. 753.)

The Affinity of Partial Agonists

A partial agonist only produces a response when a large proportion of the receptors is occupied and it should therefore antagonise the effects of a more active agonist when they are present together. If it is very feeble as an agonist, it may be possible to test the partial agonist just as if it were an antagonist, i.e. to produce measurable dose-ratios for a strong agonist with concentrations of the partial agonist which themselves have negligible agonist effects. The dose-ratios should vary with the concentration of partial agonist according to the Gaddum-Schild equation (page 158) and the affinity constant can be calculated.

If the partial agonist has more marked agonist activity its affinity may be estimated by observing the extent to which it antagonises the effects of a strong agonist in an experiment in which the effects of a concentration P of the partial agonist are matched by a concentration A_1 of a strong agonist, and a concentration A_2 is

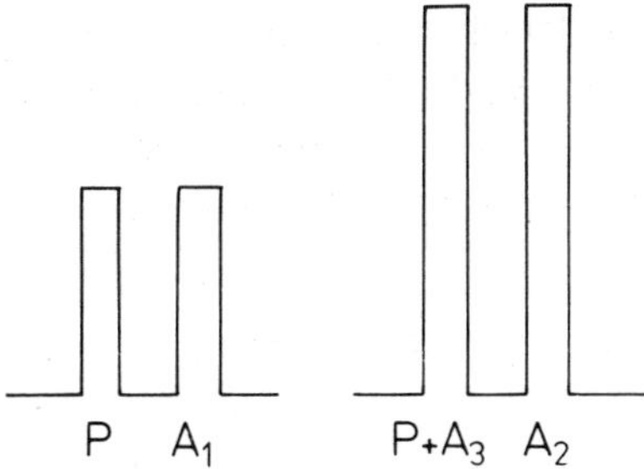

FIGURE 4.11 Addition method for the measurement of the affinity of a partial agonist. Responses produced by P of a partial agonist are matched by A_1 of agonist and those produced by $P+A_3$ are matched by A_2. In practice A_1 and A_2 are calculated by interpolation on a log dose response line.

found which matches the concentration P of partial agonist plus a concentration A_3 of the strong agonist (Figure 4.11). In practice the concentrations A_1 and A_2 are calculated from the log dose response curve, i.e. control doses of the strong agonist are given which produce responses which are likely to be roughly similar to those expected from the partial agonist alone and from the combination of partial agonist and full agonist. The concentrations of full agonist which would exactly match the (mean) response to P and the (mean) response to $P+A_3$ are then calculated from the response to the controls (page 15).

For matching responses the stimuli should be equal so $e_a y = e_p p$ where e_a is the efficacy of the strong agonist, which occupies a proportion y of the receptor and e_p is the efficacy of the partial agonist, which occupies a proportion p. If the strong agonist only occupies a small proportion of the receptors, $y = AK_a$ where A is the concentration of agonist and K_a is its affinity constant, so $e_p p = e_a K_a A_1$. If the proportion of receptors occupied by agonist is small the stimulus contributed by A_3 in the presence of P is $e_a K_a A_3 (1-p)$—strictly the proportion of receptors occupied by agonist in this situation is $A_3 K_a (1-p)/(1+A_3 K_a)$—so $e_a K_a A_2 = e_p p + e_a K_a A_3 (1-p)$ and $1-p = (A_2 - A_1)/A_3$. When p is known, K_p can be calculated because $PK_p = p/(1-p)$.

For example, Stephenson[11] obtained responses with 64 μM n-octyltrimethylammonium which matched those produced by 1.48 μM n-butyltrimethylammonium (calculated from a log dose response line obtained with 1.4 and 2.8 μM butyltrimethylammonium). When 64 μM of the n-octyl compound were present together with 5 μM of the n-butyl compound the responses were calculated to be matched by 2.46 µM *n*-butyltrimethylammonium. If the *n*-butyl compound is only occupying a small proportion of receptors,

$$1 - p = \frac{2.46 - 1.48}{5} = 0.196.$$

Hence $p = 0.804$, so $PK_p = 4.10$ and

$$K_p = \frac{4.10}{64} \times 10^6 = 6.4 \times 10^4.$$

If the partial agonist is more active still, it may be possible to obtain responses with a strong agonist which match responses from several different concentrations of the partial agonist. For the strong agonist occupying a small proportion of receptors, the stimulus is $e_a A K_a$ and this should be the same as that produced by the concentration of partial agonist which it matches, namely $e_p PK_p/(1 + PK_p)$, so $e_p/e_a K_a A = (1/PK_p) + 1$, and the graph of $1/A$ against $1/P$ should be a straight line (Figure 4.12) and when $1/A = 0$, $1/P = -K_p$.

This method involves extrapolation and the type of error associated with the Lineweaver-Burk plot (page 139) and there are advantages in using a direct fit to the hyperbola

$$\frac{e_a K_a}{e_p}(A) = \frac{PK_p}{1 + PK_p}$$

as described by Parker and Waud.[17]

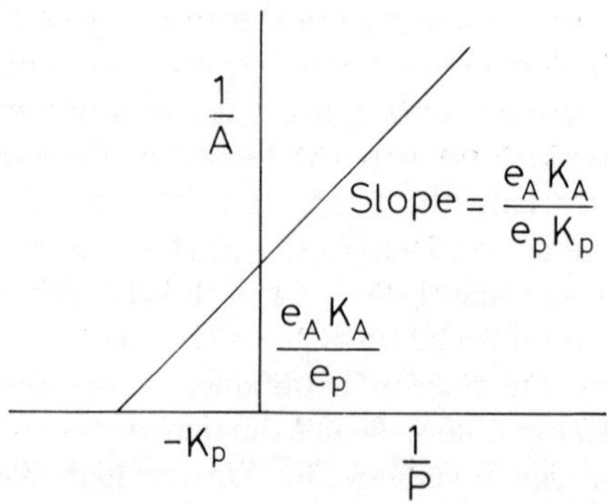

FIGURE 4.12 Reciprocal plot method for the measurement of the affinity of a partial agonist. Responses produced by P of a partial agonist are matched by A of agonist and $1/A$ is plotted against $1/P$. When $1/A = 0$, $1/P = -K_P$.

When the affinity constants of partial agonists are known it is possible to compare their efficacies. If concentrations P_1 and P_2 of two different partial agonists produce responses of the same size, $e_{p1}P_1 = e_{p2}P_2$ where e_{p1} and e_{p2} are the efficacies of the compounds and p_1 and p_2 are the proportions of receptors occupied by concentrations P_1 and P_2 of the compounds. If the affinity constants are known, p_1 and p_2 can be calculated and so it is possible to calculate the ratio of the efficacies $e_{p1}/e_{p2} = p_2/p_1$. This can also be obtained from the ratio of the concentrations producing the maximal response, i.e. when $p_1 = p_2 = 1$. It is not given by the relative sizes of the maximum responses.

From comparisons of two compounds whose affinities are known it is therefore possible to estimate the ratio of their efficacies. A scale for measuring efficacy is, however, implied by Stephenson's definition of 1 unit of stimulus as that producing half the maximum response of which the tissue is capable.[11] If the partial agonist produces such a response with a proportion, p, of receptors occupied, its efficacy is $1/p$. For responses of some other size it would be necessary to convert the size of the response into the corresponding stimulus by using the relationship between response and stimulus shown in Fig. 4.8, obtained with a full agonist.

As has already been mentioned, the shape of this curve varies greatly from one preparation to another and even on the same preparation over a period of time. This has a big effect on the 'partiality' of a partial agonist. With many preparations, particularly in practical exercises for students, the weaker partial agonists, i.e. the *n*-nonyl and *n*-octyl, and even occasionally the *n*-heptyl, trimethylammonium compounds can appear virtually inactive but when their affinity constants are measured, with the compounds treated simply as antagonists, these are found to be in the expected range. With these insensitive preparations, therefore, it is not the binding of the drug which has been affected but the link between receptor activation and tissue response.

What determines whether a compound appears to be a partial agonist is how its efficacy compares with the stimulus needed to produce a maximum response. For the particular tissue shown in Figure 4.8 it appears that this is about 8 units. A compound with an efficacy of about 8 will therefore be a full agonist. In fact, a compound with an efficacy of 5.5, producing 95% of the maximum response might appear to be a full agonist. If the response were directly proportional to receptor occupancy, however, the ratio of the concentrations producing 95% response (occupancy) to 50% response (occupancy) is

$$\left(\frac{0.95}{0.05}\right)\bigg/\left(\frac{0.5}{0.5}\right) = 19.$$

A compound with an efficacy of 8 would produce a response which was $\frac{8}{9} = 89\%$ of the maximum and one with an efficacy of 5.5 would produce $5.5/6.5 = 84\%$ of the maximum; these should clearly be partial agonists. Partial agonists should therefore be more common with tissues where the relation between response and stimulus is flatter and nearer to that determined simply by the change in occupancy.

It follows as a corollary that if a compound has high efficacy it will produce the stimulus necessary for a maximum response with only a small proportion of receptors occupied. If 8 units are needed, a compound with an efficacy of 80 could produce this with $y = 0.1$; with an efficacy of 800, $y = 0.01$ for a maximum response. The proportion needed can be calculated from the effects of an irreversible antagonist (page 162).

Example 4:10.

In experiments on the frog rectus preparation β-pyridylmethyl-trimethylammonium was a full agonist and p-aminobenzyldi-ethylamine was a partial agonist. Mean responses to 1.6×10^{-6} and 3.2×10^{-6} M agonist, recorded with a kymograph, were 37 and 53 mm, respectively. The mean response to 5×10^{-4} M partial agonist was 26 mm and when this was given together with 3.2×10^{-6} M agonist the mean response was 52 mm. Calculate the proportion of receptors occupied by this concentration of the partial agonist and its affinity constant.

(From results of Thompson, PhD thesis (1968), University of Edinburgh, Figure 5.)

Example 4:11.

In experiments on the frog rectus preparation the following responses were obtained with a full agonist (tetramethyl-ammonium) and a partial agonist (methylpyridinium):

Agonist concentration	Response	Partial agonist concentration	Response
1×10^{-5} M	4	3×10^{-3} M	6
2	8	6	13
4	26	12	22
		24	27

Calculate the concentrations of agonist which produce responses which would match those obtained with the concentrations of partial agonist and hence calculate the affinity constant of the partial agonist. If the response obtained with 4×10^{-5} M tetramethylammonium is half the maximum of which the tissue is capable, what is the efficacy of the partial agonist?

(From results of Barlow, Scott and Stephenson, *British Journal Pharmacol.* (1967), no. 31, pp. 188–96.)

The Affinity of Agonists

The relative activities of agonists can be estimated by comparing concentrations producing comparable responses but this gives no indication whether one drug fits a receptor better than another (page 30). If a concentration A_1 of an agonist with an efficacy e_{a1} and an affinity constant K_{a1} produces the same response as a concentration A_2 of a second agonist whose efficacy is e_{a2} and whose affinity is K_{a2}, the stimuli should be the same and if both occupy only a small proportion of the receptors, $e_{a1}K_{a1}A_1 = e_{a2}K_{a2}A_2$. The ratio A_1/A_2 gives no indication of the ratio of affinities or of efficacies.

It may be possible to measure the affinity of an agonist, however, by using an irreversible blocking agent to reduce the proportion of receptors available to the agonist.[17] If enough blocking agent is used the log dose response curve should become flatter and it will be no longer possible to obtain a response which is the maximum the tissue can produce (Figure 4.10). If the response produced by a concentration A of agonist by itself is also produced by a concentration A' in conditions where the irreversible blocking agent occupies a proportion z of the receptors, so only $1 - z$ are available to the agonist, the stimuli should be the same and

$$\frac{e_a K_a A}{1 + K_a A} = \frac{e_a K_a A'(1 - z)}{1 + K_a A'};$$

taking reciprocals,

$$\frac{1}{A} + K_a = \frac{1}{A'(1 - z)} + \frac{K_a}{1 - z}$$

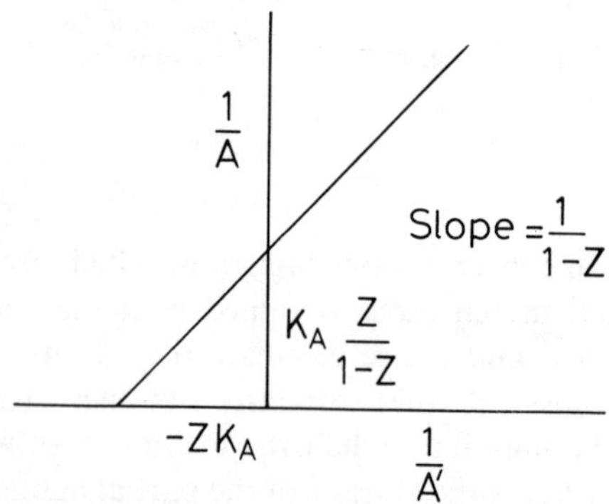

FIGURE 4.13 Reciprocal plot method for the measurement of the affinity of a full agonist. Responses produced by A' of the agonist in the presence of sufficient irreversible antagonist to flatten the log dose response curve are matched by A of agonist alone and $1/A'$ is plotted against $1/A$. The line has a slope of $1/(1-z)$ and when $1/A = 0$, $1/A' = -zK_a$.

and

$$\frac{1}{A} = \frac{1}{1-z}\left(\frac{1}{A'}\right) + \frac{zK_a}{1-z}$$

so the graph of $1/A$ against $1/A'$ should be a straight line with a slope of $1/(1-z)$ (Figure 4.13) and when $1/A = 0$, $1/A' = -zK_a$, so z and K_a can be calculated.

This again involves an extrapolation which can be avoided[18] by direct fit of A and A' to a hyperbola, $A = M(A'/(A'+K))$. The equation relating the stimuli of matching responses can be rewritten

$$A = \frac{A'(1-z)}{1+zA'K_a} = \left(\frac{1-z}{zK_a}\right)\left(\frac{A'}{A'+(1/zK_a)}\right),$$

M will be $(1-z)/zK_a$ and K will be $1/zK_a$.

Rate Theory

It has been assumed so far that the effects of an agonist or partial agonist depend on the proportion of receptors occupied. This is an extension of simple chemical ideas about the rates of catalysed reactions. Many highly active substances such as acetylcholine, are small molecules, however, and chemical changes which make them bigger usually increase affinity but reduce their activity. High affinity does not appear, in fact, to be associated with high efficacy[11] but rather with antagonism and Paton[19] has suggested that it is the rate of combination of agonist with receptor which determines the response rather than the proportion of receptors occupied by it, i.e. the response depends on 'rate' rather than 'occupancy'.

The rate of occupation of receptors $dy/dt = k_{+1}A(1-y)$, where A is the concentration of agonist, y is the proportion of receptors occupied and k_{+1} is the rate constant and at equilibrium this will equal the rate of breakdown of drug-receptor complexes, $k_{-1}y$, and $y = AK/(1+AK)$, where K is the affinity constant (page 151) so

$$\frac{dy}{dt} = k_{-1}\left(\frac{AK}{1+AK}\right) = \frac{k_{+1}A}{1+AK}.$$

This is a hyperbola and the expression is indistinguishable mathematically from the corresponding equation for occupation theory—$y = AK/(1+AK)$. If the rate of occupation determines the stimulus, however, a rate constant, k_{+1} or k_{-1}, might be regarded as replacing the term 'efficacy' used in occupation theory.

Although it is not possible to distinguish between the two theories at equilibrium, it might be possible to do so by studying events before equilibrium is reached. The response of a piece of guinea-pig ileum to acetylcholine often rises to a peak and then falls to a lower level, a phenomenon described as 'fade' and it was suggested that this corresponded to the initial fast rate of combination of agonist with receptor followed by a slowing as equilibrium was reached. The proportion y of receptors occupied at a time t is related to the proportion y_e occupied at equilibrium by the equation $y_t = y_e(1 - e^{-(k_{+1}A + k_{-1})t})$ which can be written $y_t = y_e(1 - e^{-k^*t})$ whereas the rate of occupation dy/dt is:

$$k_{+1}A(1 - y_t) = k_{+1}A[1 - y_e(1 - e^{-k^*t})]$$
$$= k_{+1}A(1 - y_e + y_e e^{-k^*t})$$
$$= y_e(k_{-1} + k_{+1}A e^{-k^*t})$$

because $y_e = AK/(1+AK)$ and $1 - y_e = y_e/AK$.

The first expression indicates that occupancy rises gradually to a maximum; the second expression indicates that the rate of occupancy starts at a high value ($t = 0$) and then declines to a steady level, which is much more consistent with fade. Not all tissues, or even preparations from the same tissue, show fade, however, and it is very difficult to establish that it is really a phenomenon involving drug and receptor rather than one involving the mechanism of the contractile response or perhaps arising from problems of diffusion. When the drug is washed out, the proportion of receptors occupied should decline exponentially and $y_t = y_e e^{-k_{-1}t}$, but the rate of occupation should drop to zero, provided the washing process is effective and there is really no agonist present. Studies have been made of the rates of onset and offset of antagonists, which are much slower and can be measured more easily, and these indicate that the rates observed are at least partly limited by diffusion.[20]

There is no convincing evidence either for or against the idea that the response is determined by the rate of combination of drug and receptor but it is remarkable that antagonists never produce any obvious stimulation in conditions in which they are combining quite fast with receptors. It has been suggested that perhaps it is the rate of breakdown of the complex, rather than its rate of formation, which determines the response, i.e. k_{+1} should be replaced by k_{-1}. The rate of breakdown is known to be slow for antagonists. In this form, however, the theory is still mathematically indistinguishable from occupation theory at equilibrium. In either form, 'onset rate' or 'offset rate', the two rate constants k_{+1} and k_{-1} would determine the activity of an antagonist and there would be no need to introduce the concept of efficacy.

Rate theory would explain why increased binding leads to decreased agonist activity but there are other explanations for this once it is appreciated that the receptor itself may undergo a change.

EXAMPLE 4:12.

The table shows the dose-ratios produced during the onset and offset of the antagonism of acetylcholine by methylatropinium $(1.5 \times 10^{-9} \, \text{M})$ on the guinea-pig ileum:

Time after adding methylatropinium (onset) or washing it out (offset)	Dose-ratios	
(min)	Offset	Onset
1	4.54	1.28
2	3.23	1.67
3	2.70	2.33
4	2.38	3.03
5	2.13	3.70
6	2.00	4.55
7	1.92	5.26
8	1.72	
9	1.64	
10	1.56	

The dose-ratio at equilibrium was 6.67. Use the dose-ratios to calculate the proportion of receptors occupied by antagonist and hence the rate constants for offset and onset of antagonism (do the calculations in this order). Compare the equilibrium constant calculated from the ratio of the rate constants with the value obtained from the equilibrium dose-ratio.

(From results of Paton and Rang, *Proc. Roy. Soc. B* (1965), no. 163, Figure 1, p. 8.)

EXAMPLE 4:13.

The dose of atropine producing detectable effects in man is about $0.02\,mg\,kg^{-1}$. It is distributed throughout the total body fluid. Calculate the approximate molar concentration in the blood-stream (the molecular weight is 347). The affinity constant for atropine and muscarinic acetylcholine receptors is approximately $1 \times 10^9\,l\,mol^{-1}$. What dose-ratio should this concentration produce and what proportion of receptors will be occupied by atropine?

EXAMPLE 4:14.

The dose of $(+)$-tubocurarine chloride used in man is about $10\,mg$. It is distributed in the extracellular fluid. Calculate the approximate molar concentration in the blood-stream of a subject weighing $70\,kg$ (the molecular weight is 786). The affinity constant for acetylcholine receptors in the neuromuscular junction is about $3 \times 10^6\,l\,mol^{-1}$. What dose-ratio should this concentration produce and what proportion of receptors will be occupied by $(+)$-tubocurarine?

Compare the dose-ratio with that produced by atropine in Example 4:13. Why are the doses chosen to produce effects of such different size? Compare the molar concentrations with the molar concentration of ethanol equivalent to $80\,mg\,\%$. Why is there such a difference?

Desensitisation of Receptors

The term *tachyphylaxis* has been used for many years to describe what happens when repeated applications of the same dose of a drug produce smaller and smaller responses. This can be seen, for instance, with the actions of nicotine on the blood-pressure or on a piece of intestine. The rate of the decline in response depends on the agonist, the size of the dose, and how frequently it is given. The decrease in response is not due to fatigue because it is still possible to obtain a full response with another agonist provided it acts on different receptors. A piece of guinea-pig ileum which shows tachyphylaxis to nicotine, for example, will still respond normally to histamine. The process, therefore, involves the receptors, which have become desensitised, and this is particularly striking with the acetylcholine receptors in the neuromuscular junction of voluntary muscle. Here there is a whole class of drugs (e.g. decamethonium and suxamethonium) which are agonists but nevertheless act as blocking agents because they produce desensitisation extremely rapidly. Desensitisation can also be produced by the natural transmitter, acetylcholine, if it is allowed to accumulate by preventing its hydrolysis by acetylcholinesterase.

Desensitisation of receptors must involve some physical change in them which can occur in various ways. Katz and Thesleff[21] have considered several models and the simplest is:

$$A+R \; \rightleftharpoons \; AR \; \underset{k_2}{\overset{k_1}{\rightleftharpoons}} \; AR'$$
$$\text{(fast)} \qquad \text{(slow)}$$

where AR′ represents the complex with the desensitised receptor. The affinity constant

$$\alpha = \frac{AR}{A(1-AR-AR')} \quad \text{and} \quad AR = \frac{\alpha A(1-AR')}{1+\alpha A}.$$

If the forward rate constant for the desensitising step is k_1 and the backward rate constant is k_2,

$$\frac{AR'}{AR} = \frac{k_1}{k_2},$$

and the desensitised fraction,

$$AR' = \frac{k_1}{k_2}\left[\frac{\alpha A(1-AR')}{1+\alpha A}\right] = \frac{1}{1 + \dfrac{k_2}{k_1}\left[\dfrac{1+\alpha A}{\alpha A}\right]}.$$

With this model the desensitised receptor R′ only exists as a complex but in the following cyclical model it can exist independently:

$$A + R \; \rightleftharpoons \; AR \dashrightarrow \text{response}$$

$$(\text{slow}) \quad k_2 \!\uparrow\!\downarrow k_4 \quad k_3 \!\uparrow\!\downarrow k_1 \quad (\text{slow})$$

$$A + R' \; \rightleftharpoons \; AR'$$

The proportion of receptors in the desensitised form can be calculated to be

$$\cfrac{1}{1 + \cfrac{k_2(1 + \alpha A)}{k_4(1 + \beta A)}}$$

where A is the concentration of agonist, α is the association constant of agonist and active receptor, β is the association constant of agonist and desensitised receptor, k_2 is the rate constant for the conversion of desensitised receptor back to active receptor and k_4 is the rate constant for the conversion of active receptor into desensitised receptor. It is assumed that processes involving changes in the receptor ($R \rightleftharpoons R'$ and $AR \rightleftharpoons AR'$) are slow compared with those involving drug and receptor ($A + R \rightleftharpoons AR$ and $A + R' \rightleftharpoons AR'$). This model fits experimental results provided $\beta/\alpha > k_1/k_2$ and suggests that even in the absence of agonists there is an equilibrium between active and desensitised receptors.

It is not entirely certain that desensitisation is a receptor phenomenon, rather than an effect on closely following steps in the activation process.[22] The idea that it is a receptor process, however, provides a relatively simple explanation for the 'metaphilic' effect.[23] Alkylating agents have been developed with considerable specificity for the ('nicotinic') acetylcholine receptors in the neuromuscular junction and in slowly contracting muscle such as the *biventer cervicis* of the chick. They react with the receptors and block them irreversibly. It was found, however, that the extent of the block was greater when the treatment with the antagonist was preceded by exposure of the tissue to agonist. It appears that the alkylating agents have a bigger affinity for the desensitised form of the receptor and that exposure to the agonist increases the proportion of desensitised receptors AR', which can react with the alkylating agent when they have dissociated into $A + R'$. This increase in the activity of the antagonist brought about by the agonist has been called the metaphilic effect. It has also been found that recovery from a desensitising dose appears to be a first order process which is independent of the agent used to desensitise the receptors. The rate constants for the onset of desensitisation, however, are different for different drugs.

Dynamic Receptors and Allosteric Effects

Because of the results obtained in studies of desensitisation and because of the discovery of allosteric mechanisms in enzymes it is probably necessary to abandon the idea that receptors can be represented as a single species (R) in their interaction with drugs (A). The equation

$$A + R \rightleftharpoons AR$$

is almost certainly an oversimplification. It is, however, difficult to obtain definite evidence for allosteric behaviour by receptors because of the large experimental errors attached to most results. When these are examined by a Hill plot their variance is usually such that it is impossible to establish that the slope is significantly different from 1.

In many situations, particularly with electrophysiological responses, it is not possible to determine the maximum response so a Hill plot cannot be made. If, however, accurate results can be obtained with low concentrations, the graph of response against concentration (*not* log concentration) may be inspected to see if it

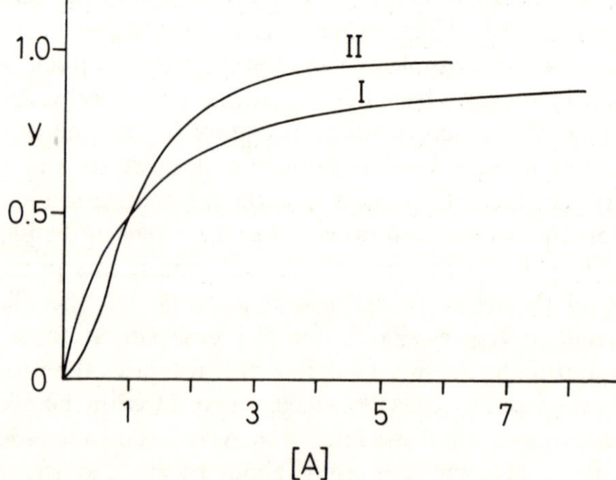

FIGURE 4.14 Curve I is the graph of $y = AK/(1 + AK)$:
  Curve II is that of $y = A^2K/(1 + A^2K)$: with both graphs $K = 1$. Note the slight sigmoid shape of curve II (see Example 4:2).

is sigmoid and departs from the Langmuir isotherm (Figure 4.14). This was observed by Katz and Thesleff[21] for the depolarisation of the frog neuromuscular junction produced by carbachol. Sigmoidicity has also been found with the effects of γ-aminobutyric acid on the membrane conductance in the neuromuscular junction of the crayfish[24] and in other systems (reviewed by Colquhoun[25]). With some of these electrical events, particularly changes in membrane conductance, it seems distinctly possible that the response may be directly proportional to receptor occupancy and, provided this is so, the results indicate that the binding does not follow the simple Langmuir model.

Further evidence for the need to abandon the simple model comes from studies of the binding of agonists and antagonists to preparations of receptor material. Certain parts of the brain are rich in muscarine-sensitive acetylcholine receptors and it is easy to measure the affinities of compounds for isolated preparations of membrane fragments containing these receptors (e.g. Example 4:4). The affinity constants of antagonists are very similar to values in functional tissues, such as the isolated guinea-pig ileum,[26] suggesting that the process of breaking up the cells has not affected receptor structure and that the receptors in the two tissues are very similar, even though one type comes from cells in rat brain and the other from cells in guinea-pig intestine. This similarity can be seen in Figure 4.15 in which estimates of log affinity constant obtained in experiments with the guinea-pig ileum as described on pp. 32, 158 are compared with estimates of log affinity constant obtained with the membrane fragments by competition with the binding of a labelled compound as described on page 133. If the receptors were exactly the same the value of $\log K$ for a compound should be the same and the results for a series of compounds should lie along the line of identity. The agreement in Figure 4.15 is all the more remarkable because in addition to the difference in source and species, measurements on the ileum were made at 37°C and those with the fragments were made at 30°C.

The binding of these antagonists appears to be satisfactorily described by the Langmuir isotherm; Hill plots all have a slope close to 1. With agonists, however, there is evidence for a second binding site on the membrane fragments.[27] It is not known how this may be related to the action of agonists but clearly it is necessary to consider how results obtained in the various tests used to measure the affinity of antagonists, partial agonists and agonists

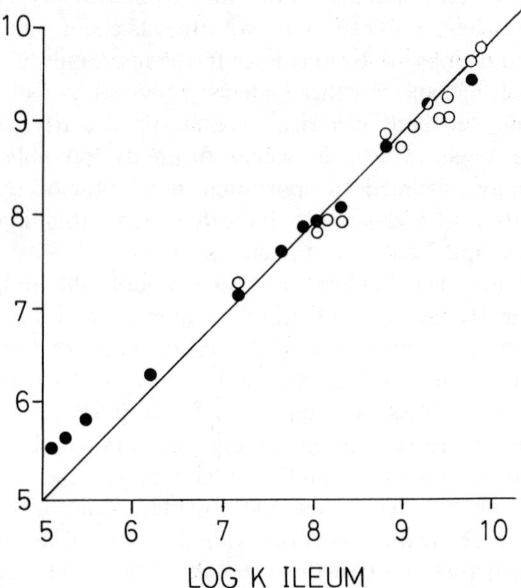

FIGURE 4.15 Investigation of receptor identity. Estimates of log affinity constant of a compound for 'muscarinic' acetylcholine receptors on membrane fragments obtained from rat brain are plotted against estimates for functional receptors on the isolated guinea-pig ileum. If the binding sites are identical, the values of $\log K$ should be the same and the results for different compounds should lie along the line of identity. The binding to the membrane fragments was measured by competition with (^{3}H)-*N*-methylatropinium (O) or with (^{3}H)-propylbenziloylcholine (●) in experiments at 30°C. The results with the guinea-pig ileum were obtained from measurements of dose-ratios and were made at 37°C. It appears that the two types of receptors are very similar and that breaking up the cell has not seriously altered the structure, at least so far as the binding of antagonists is concerned.

(Drawn from results of Hulme, Birdsall, Burgen and Mehta, *Mol. Pharmacol.* (1978), no. 14, p. 737.)

may be affected if it is necessary to use a two-site or allosteric model. The problem has been reviewed comprehensively by Colquhoun.[25]

Estimates of Affinity in Allosteric Systems

According to the model proposed by Monod, Wyman and Changeux,[28] the binding protein is made up of n subunits and can exist in two forms, T and R. The ligand, A, can bind up to n molecules with either form and the equilibrium constants for each step can be written K_{Ti} or K_{Ri} where these are the dissociation constants for $T_nA_i \rightleftharpoons T_nA_{(i-1)}$ and $R_nA_i \rightleftharpoons R_nA_{i(i-1)}$. The ratio $K_{Ri}/K_{Ti} = M_i$, which indicates the preference shown by the ligand for the two forms. If there is stronger binding to the R form the dissociation constant will be less and $M < 1$. The equilibrium constant for the binding protein in the absence of drug is $L = [T_n]/[R_n]$ and it is convenient to use the 'normalised' concentration of the ligand, $c = x/K_R$, where x is the actual concentration of ligand and K_R is the dissociation constant of the complex R_nA (c is equivalent to AK or S/K_s on pp. 138, 151). The situation can be represented:

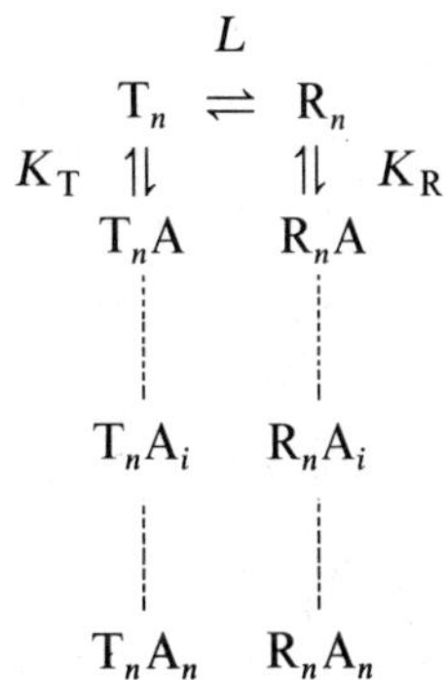

If the R form is regarded as the active form (perhaps associated with open ion channels in a membrane) the proportion of active form:

$$P_R = \cfrac{1}{1 + L\left(\cfrac{1 + \Sigma Mc}{1 + \Sigma c}\right)^n} .$$

This can be compared with the proportion

$$y = \cfrac{1}{1 + \cfrac{1}{c}}$$

for the single receptor model written in the same way (page 151).

The equations for two drugs, such as a partial agonist and a full agonist or an antagonist and a full agonist are more complex but it appears that with the method used for measuring the affinity of antagonists from dose-ratios (pp. 32, 158)

$$K_{est} = \frac{(1 - M_A)}{(1 - M_A/M_B) + (1 - 1/M_B)(K_{AR}/x_A)} K_{BT}$$

where K_{est} is the dissociation constant estimated experimentally, K_{BT} is the constant for the dissociation of the complex of the antagonist with the inactive form, K_{AR} that for the complex of the agonist with the active form, x_A is the normalised concentration of agonist used to produce the response in the absence of antagonist and M_A and M_B indicate the selectivity of the agonist and antagonist for the two forms. If the antagonist has an equal preference and $M_B = 1$, the experimentally estimated value will be the same as K_{BT}. If it has greater preference for the inactive form ($M_B > 1$), the results will still fit the Gaddum-Schild equation provided the dose-ratios are all measured at the same response level and x_A is constant. The results will be affected by the choice of agonist used and for K_{est} to be close to K_{BT} it is desirable to use an agonist with as great a preference as possible for the active form ($M_A < 1$ and $M_A \ll M_B$) and also with a high affinity for it, so the dissociation constant K_{AR} is small.

In the estimation of the affinity of partial agonists by comparing concentrations of partial agonist and full agonist producing comparable responses (page 166) the dissociation constant

$$K_{est} = K_{BT}\frac{(1 - M_A)}{(1 - M_A/M_B)}$$

so it is again desirable to use an agonist which is selective for the active form. Provided M_A is small this method should give a fairly close estimate of K_{BT}, the dissociation constant for the complex formed by the partial agonist and the inactive form of the receptor.

For the estimation of the affinity of full agonists the situation is much more complex and Colquhoun writes 'the present treatment suggests that there is at present not enough knowledge to interpret with any certainty the irreversible antagonist method in terms of cooperative models'.

It seems, therefore, that it is possible to measure the binding of competitive antagonists and possibly of partial agonists to receptors on intact cells. The binding of agonists, however, is difficult to measure.

Forces between Drug and Receptor

The biological properties of a drug depend on its interaction with cells. At the molecular level this involves effects on membranes and/or effects on receptors and/or effects on enzymes. The forces between drug and binding site in receptor or enzyme are now briefly reviewed.

Some drugs react with receptors forming a covalent bond and their effects are irreversible. Many of these are β-haloethylamines containing the group

$$\diagdown N CH_2CH_2X,$$

e.g. SY-28 (page 161). These are stable when they are salts but once the free base is liberated the halogen reacts with the tertiary nitrogen atom forming an aziridinium ('immonium') ion. This rearranges with the opening of the ring to form a carbonium ion which is highly reactive with any group having an excess of electrons, such as $HO-$, $HS-$, $HN=$:

In water $R'' = H$ and the alcohol is formed, so the biological action of these compounds depends upon the binding group in the receptor or enzyme being appreciably more nucleophilic than water.

The 'nitrogen mustards', used in the treatment of some forms of malignant disease, contain two β-haloethylamine groups, e.g. Mustine, $MeN(CH_2CH_2Cl)_2$, HCl. Other examples of substances which form covalent bonds with their site of action are the fluophosphonates, e.g. Dyflos (DFP) or tetraethylpyrophosphate (TEPP); these also react with groups with an excess of electrons and lose a negative ion:

$$\begin{array}{ccc}
iso\mathrm{PrO} & & iso\mathrm{PrO} \\
\diagdown & & \diagdown \\
iso\mathrm{PrO}\!-\!\mathrm{P}\!\!=\!\!\mathrm{O} + \mathrm{HOR''} \rightarrow & iso\mathrm{PrO}\!-\!\mathrm{P}\!\!=\!\!\mathrm{O} \\
\diagup & & \diagup \\
\mathrm{F} & & \mathrm{R''O} + \mathrm{H^+} + \mathrm{F^-}
\end{array}$$

Although such highly reactive groups make for strong binding, this may not be specific for a particular receptor and the compounds may have undesirable effects because they act at many sites. The specificity will be determined by the groups R and R' and with the need to include the reactive part ($-CH_2CH_2Cl$ or F), it may be difficult to produce a really specific drug.

Many drugs produce effects which are reversible and the binding involves much weaker forces, though with large molecules the combination of weak forces can lead to very high affinity. If the receptors or enzymes are structures with quite individual geometry, it should be possible to produce compounds which have high specificity as well as high affinity, because pharmacodynamic groups in the drug are so arranged that there are many relatively weak interactions with complementary groups in the receptor or enzyme.

Among the strongest of these interactions should be the electrostatic attraction between ions of opposite charge. This has been calculated to contribute about $5\,\mathrm{kcal\,mol^{-1}}$ ($21\,\mathrm{kJ\,mol^{-1}}$) to the Gibbs free energy and the corresponding increase in the value of $\log K$, calculated from the van't Hoff relation is 3.6 log units. A compound able to form such a bond ought therefore to have 4000 times the affinity of one which could not do so. There may be interactions between drug and receptor involving smaller units of charge and contributing correspondingly less to the stability of the drug-receptor complex. Hydrogen bonds have been estimated to make very similar contributions—about 5 kcal mol^{-1} (21 kJ mol^{-1})—though these bonds vary considerably in strength, depending on steric factors as well as on the nature of the groups attached to hydrogen. Van der Waals' forces (page 84) are usually regarded as weak, contributing perhaps 0.5 kcal mol^{-1} (2.1 kJ mol^{-1}), and they decrease markedly with distance (being proportional to $1/d^7$), but might make a significant contribution if there are many points of contact. The values quoted for these forces are based largely on calculations by Pauling,[29] which were summarised by Albert,[30] but they are likely to need modification to take into account the effects of water.

Water is a highly associated liquid (for reviews see ref. 31). It has been calculated that only about 12% of the hydrogen bonds in ice are broken by melting; at 40°C about 75% are still intact and even at 100°C over 50% are intact, which is why water has so high a boiling point. According to one model, water can be regarded as being made up of 'flickering' crystals, aggregates which form and break up with a life of about 10^{-11} s and about 40 molecules/aggregate at 40°C. Inert gases dissolve in water to some extent and surprisingly the process is exothermic. The limited solubility of the compounds indicates that ΔG is negative but small, and as ΔH is also appreciably negative, ΔS must also be negative because $\Delta G = \Delta H - T \Delta S$.

Solution would be expected to be accompanied by an increase in entropy (disorder) but with inert gases dissolved in water there is an increase in order. This can only come from the water, not from the inert gas, and in some instances it can be so great that crystalline compounds are formed, with the water molecules in a cage-like structure about the inert gas. These are termed clathrate hydrates and their occurrence is not confined to inert gases. They can be formed by a variety of non-polar compounds, including some general anaesthetics.

The ordering effect of non-polar groups on water is termed the hydrophobic effect and is of great importance in determining the conformation of macromolecules in water. Aliphatic side-chains in the amino acids of a protein, for example, tend to come together in the presence of water so that the increased order is minimal. They are said to be 'hydrophobically bonded' but this does not mean that there is an attraction between them as would occur with van der Waals' forces; the attraction depends on the presence of the water and it does not require the very close fit which would be necessary for van der Waals' forces to be appreciable. In fact, the contribution of van der Waals' forces to the binding of drug with receptor seems likely to be very much weaker than contributions due to hydrophobic interactions, which have been estimated to be about $0.75\,\mathrm{kcal\,mol^{-1}}$ $(3.1\,\mathrm{kJ\,mol^{-1}})$ for one methylene group $(-CH_2-)$.

It might be expected that ions would have an ordering effect on water because the ionic charge should produce a shell of water molecules around it with the lone pairs of electrons on the oxygen attracted towards positive charges, or with the hydrogen atoms oriented towards negative charges. In producing this local order around the ion, however, there is extensive destruction of the overall order of water. In contrast to hydrophobic interactions which are structure-making, ionic interactions are likely to be structure-breaking, though there is no sharp division of substances into one class or the other. The contribution to the binding of drug to receptor from the attraction between ions of opposite charge however, is likely to be much less when these are hydrated than when they are naked because association would be accompanied by an overall increase in order (decrease in entropy).

The same considerations will apply to the formation of hydrogen bonds between drug and receptor. It seems most unlikely that in a strongly hydrogen-bonded solvent such as water a drug could form hydrogen bonds with a receptor comparable with those observed in crystals, for instance.

Drug action at the molecular level involves processes related to adsorption, such as those considered in this chapter. What happens depends on the forces between the drug and the receptor or enzyme and these are determined largely by the size and shape of the drug molecule and the geometry of the binding site. It is not clear how far our ideas about the binding forces need to be modified to allow for the effects of water. It might be expected that the receptor must be in an environment which allows water-soluble drugs access, but there are areas in many proteins which are relatively occluded from water. The strength of the forces between drug and receptor may therefore depend on how much water there is in the region where the drug is being bound. There is, then, some uncertainty as to what contribution to binding can be expected from pharmacodynamic groups and it is necessary to find out experimentally by studying the effects of chemical structure on affinity. This is considered in the next chapter, which deals with the possibility of predicting drug activity.

References

1. I. Langmuir, *Journal American Chem. Soc.* (1916), no. 38, p. 2221; ibid. (1918), no. 40, p. 1361.

2. A. V. Hill, *Biochem. Journal* (1913), no. 7, p. 471; *Biochem. Journal* (1921), no. 15, p. 577; W. E. L. Brown and A. V. Hill, *Proc. Roy. Soc. B.* (1923), no. 94, p. 297.

3. G. Scatchard, *Ann. NY Acad. Sci.* (1949), no. 51, p. 660.

4. I. M. Klotz, J. M. Urquhart and W. W. Weber, *Arch. Biochem.* (1950), no. 26, p. 427.

5. E. De Robertis, G. S. Lunt and J. L. La Torre, *Mol. Pharmacol.* (1971), no. I, p. 97.

6. H. Lineweaver and D. Burk, *Journal American Chem. Soc.* (1934), no. 56, p. 658.

7. K. B. Augustinsson, *Acta Physiol. Scand.* (1948), no. 15, suppl., no. 52, p. 100.

8. B. H. J. Hofstee, *Science* (1952), no. 116, p. 329.

9. A. J. Clark, 'General Pharmacology', *Handbuch der Experimentellen Pharmakologie*, IV (Springer, Berlin, 1937), p. 64.

10. E. J. Ariëns and J. M. van Rossum, *Arch. int. Pharmacodyn.* (1957), no. 110, p. 275.

11. R. P. Stephenson, *British Journal Pharmacol.* (1956), no. 11, p. 379.

12. J. H. Gaddum, *Journal Physiol.* (1937), no. 89, p. 7P.

13. H. O. Schild, *British Journal Pharmacol.* (1947), no. 2, p. 189; ibid. (1949), no. 4, p. 277.

14. O. Arunlakshana and H. O. Schild, *British Journal Pharmacol.* (1959), no. 14, p. 48.

15. M. Nickerson, *Nature* (1956), no. 178, p. 697.

16. J. H. Gaddum, *Trans. Faraday Soc.* (1943), no. 39, p. 323; *Pharmacol. Review* (1957), no. 9, p. 211.

17. R. F. Furchgott, in *Advances in Drug Research* 3, N. J. Harper and A. B. Simmonds (eds.) (Academic Press, London, 1966), p. 21.

18. R. B. Parker and D. R. Waud, *Journal Pharmacol.* (1971), no. 177, p. 1.

19. W. D. M. Paton, *Proc. Roy. Soc. B.* (1961), no. 154, p. 21.

20. F. Roberts and R. P. Stephenson, *British Journal Pharmacol.* (1976), no. 58, p. 57.

21. B. Katz and S. Thesleff, *Journal Physiol.* (1957), no. 138, p. 63.

22. L. G. Magazanik and F. Vyskocil, in *Drug Receptors*, H. P. Rang (ed.) (Macmillan, London, 1973), p. 105.

23. H. P. Rang and J. M. Ritter, *Mol. Pharmacol.* (1969), no. 5, p. 394; ibid. (1970), no. 6, pp. 357, 383; H. P. Rang, *British Journal Pharmacol.* (1973), no. 48, p. 475.

24. A. Takeuchi and N. Takeuchi, *Journal Physiol.* (1969), no. 205, p. 377.

25. D. Colquhoun, in *Drug Receptors*, H. P. Rang (ed.) (Macmillan, London, 1973), p. 149.

26. E. C. Hulme, N. J. M. Birdsall, A. S. V. Burgen and P. Mehta, *Mol. Pharmacol.* (1978), no. 14, p. 737.

27. N. J. M. Birdsall, A. S. V. Burgen and E. C. Hulme, *Mol. Pharmacol.* (1978), no. 14, p. 723.

28. J. Monod, J. Wyman and J-P. Changeux, *Journal Mol. Biol.* (1965), no. 12, p. 88.

29. L. Pauling, *The Nature of the Chemical Bond* (Cornell University Press, Ithaca, New York, 1939); ibid., 3rd edn (1960).

30. A. Albert, *Selective Toxicity*, 1st edn (Methuen, London, 1951), Table 1, p. 26; ibid., 5th edn (1973), p. 222.

31. F. Franks, *Chemistry and Industry* (1968), p. 560; M. J. Tait and F. Franks, *Nature* (1971), no. 230, p. 91; C. J. Tanford, *The Hydrophobic Effect* (Wiley, New York, 1973); A. Suggett, in *Biological Activity and Chemical Structure*, J. A. Keverling-Buisman (ed.) (Elsevier, Amsterdam, 1977), p. 95.

CHAPTER 5

The Prediction of Drug Activity

'It will probably always be more important to try a thing out than to argue about it'. J. H. Gaddum, *Journal Pharm. Pharmacol.* (1954), no. 6, p. 511.

Nonspecific Activity

If the biological effects of a drug depend on some physical action on cells it would seem obvious that it should be possible to predict its biological activity from its physical properties. A biological effect is only known to be nonspecific, however, when it has been shown to be independent of chemical structure, so the prediction of activity is rather an extrapolation from what is already known than a prediction of something completely new. It may, nevertheless, be useful to know how to make drugs which are more active.

Nonspecific drugs include volatile anaesthetics and a variety of substances which cause the death of certain organisms and are used as disinfectants and pesticides. The relative potencies of the compounds can be assessed from the concentrations required to produce a particular effect, such as a steady level of anaesthesia, and the range of activity can be large. For the maintenance of anaesthesia in man about 1% (v/v) of chloroform is needed, compared with about 10% of diethylether. With nitrous oxide it is not really possible to produce full anaesthesia with concentrations which are 80%. At atmospheric pressure higher concentrations are likely to cause anoxia ($O_2 < 20\%$) but higher partial pressures can be achieved by raising the pressure above atmospheric. At 10 atmospheres even nitrogen produces anaesthesia (so deep-sea divers need to breathe a mixture of helium and oxygen), but the partial pressure of nitrogen is $7600/7.6 = 1000$ times that needed with chloroform.

At the end of the last century it was realised that there was a correlation between anaesthetic potency and partitioning into lipid-like solvents such as olive oil. Potent anaesthetics had a high lipid-solubility[1] and it seemed that the concentration of material required to produce anaesthesia was determined by what was needed to produce a particular level in lipid, which was roughly the same for many drugs, being between 0.05 and 0.1 M in olive oil (Table 5.1). For dimethylacetal, with an oil:gas partition coefficient of 100 this corresponds to a concentration of 1.9% by volume in the inspired air, so it would be expected that for a substance with an oil:gas partition coefficient of 10 the concentration should be about 19% and for one with a coefficient of 1 it should be 190%, so the pressure would have to be raised above atmospheric.

TABLE 5.1: *Narcosis of Mice.*

	P	N	C_{oil}	p_s	Activity
Nitrous oxide	1.4	100	0.06	59300	0.01
Acetylene	1.8	65.0	0.05	51700	0.01
Dimethylether	11.6	12.0	0.06	6100	0.02
Methylchloride	14.0	6.5	0.07	5900	0.01
Ethylene oxide	31.0	5.8	0.07	1900	0.02
Ethyl chloride	40.5	5.0	0.08	1780	0.02
Diethyl ether	50.0	3.4	0.07	830	0.03
Methylal	75.0	2.8	0.08	630	0.03
Ethyl bromide	95.0	1.9	0.07	725	0.02
Dimethyl acetal	100	1.9	0.06	288	0.05
Diethyl formal	120	1.0	0.05	110	0.07
CHCl=CHCl	130	0.95	0.05	450	0.02
Carbon disulphide	160	1.1	0.07	560	0.02
Chloroform	265	0.5	0.05	324	0.01

P = partition coefficient (oil:vapour); N = narcotic concentration (volume %); C_{oil} = concentration (molar) of the substance in olive oil which would be in equilibrium with the narcotic concentration; p_s = saturated vapour pressure (mm Hg) at 37°C; Activity = p_t/p_s where p_t, the partial pressure in the anaesthetic mixture is calculated by multiplying the narcotic concentration by atmospheric pressure, e.g. for diethyl formal $p_t = \frac{1}{100} \times 760$ mm Hg.

(From results of Meyer and Hemini, *Biochim. Z.* (1935), no. 277, p. 54.)

The anaesthetic must first pass into the bloodstream, however, before it can reach the brain and produce its effects. This is unlikely to affect what happens at equilibrium but it may affect the rate at which equilibrium is reached. If the drug has very little solubility in blood it may require an infinite time to produce any effect. If it is very soluble in blood it is possible that this, too, may delay equilibrium because it will be necessary to inhale a large quantity of material before the level in the lipid phase can rise appreciably and there are physiological limits to how fast the drug can be inhaled (page 53).

There is no reason for believing that anaesthetics produce their effects on a structure resembling olive oil but Ferguson has pointed out that, whatever the structure is at which the drugs act, the maintenance of anaesthesia is a situation close to equilibrium, so the chemical potential should be the same in all phases and can be measured in the anaesthetic mixture.[2]

Because free energy is an extensive property and depends on the number of moles present, it is often convenient to use the chemical potential, μ, which is the Gibbs free energy/mole (also called the partial molal free energy), i.e. $\mu = (dG/dn)_{T,p}$. The chemical potential is the same as the free energy for one mole and values of ΔG for reactions are usually expressed as changes/mole and are therefore changes in chemical potential. For n moles of a substance with a partial pressure p and a standard chemical potential of μ_0, the free energy $G = n(\mu_0 + RT \ln p)$ and the chemical potential $\mu = \mu_0 + RT \ln p$. If a drug is allowed to equilibrate between two phases and the chemical potential of the drug is not the same in each, the difference will result in the transfer of molecules from one phase to the other until the partial pressures (or concentrations) have become such that the chemical potential is the same in both. Accordingly, $\mu_{01} + RT \ln p_1 = \mu_{02} + RT \ln p_2$ where the standard chemical potentials in the two phases are μ_{01} and μ_{02} and the partial pressures are p_1 and p_2. This can be written

$$-(\mu_{02} - \mu_{01}) = RT \ln \frac{p_2}{p_1}$$

which is the van't Hoff relation (though this is usually written $-\Delta G = RT \ln K$, where ΔG is actually the change in free energy/mole, the full expression is $-\Delta G = nRT \ln K$).

The absolute value of the chemical potential in the anaesthetic mixture will be difficult to measure because the chemical potential in the standard state μ_0 would need to be known. The value relative to the standard state, however, will be $RT \ln p_t/p_s$, where p_t is the partial pressure in the anaesthetic mixture and p_s is the saturated vapour pressure, obtained with pure liquid (i.e. a mole fraction of 1). Ferguson has therefore used p_t/p_s as a measure of *thermodynamic 'activity'* and has found it to be remarkably constant for a variety of anaesthetics (Table 5.1). This indicates that the change in chemical potential in going from the standard state to the biophase where the drug acts is about the same for all anaesthetics. This does not explain how the drugs are acting but the fact that values of p_t/p_s are all about 0.02 can be used, at least empirically, to predict the concentrations likely to be needed for the production of anaesthesia by volatile compounds whose saturated vapour pressure is known. The lower the saturated vapour pressure (p_s), the lower should be the partial pressure in the anaesthetic mixture (p_t). For a compound with $p_s = 200$ mm Hg, p_t should be 4 mm Hg and the anaesthetic concentration would be $\frac{4}{760} \times 100 = 0.5\%$. The value of p_s should be related to the boiling point of the compound, where $p_s =$ atmospheric pressure (760 mm Hg); the lower the boiling point the higher should be p_s. This can therefore be used as a rough prediction of anaesthetic potency. The higher the boiling point the lower should be p_s and the more active the compound is likely to be, but there must somewhere be an upper limit where the compound becomes too involatile to be inhaled.

Ferguson made similar calculations with the activities of many other compounds in a variety of biological tests, mostly for toxic effects, e.g. on bacteria, sea urchin eggs, red spiders and wireworms. It is arguable how far thermodynamic ideas based on equilibria can be applicable to toxicity, which is irreversible, but it seems clear that the activities of the compounds in these tests depend on their physical properties and will be predictable from them, if only empirically.

EXAMPLE 5:1.

Use the values in Table 5.1 to plot the graphs of log narcotic concentration against log partition coefficient; log narcotic concentration against log saturated vapour pressure; and log partition coefficient against log saturated vapour pressure. If you have access to a suitable calculator or computer, calculate the regression coefficients and compare them.

EXAMPLE 5:2.

The boiling point of F_3CCHBr_2 is 73°C, the saturated vapour pressure at 20°C is 104 mm Hg, and the anaesthetic concentration in mice is 0.4%. Calculate the thermodynamic activity (p_t/p_s). The boiling point of halothane, $F_3CCHClBr$, is 50°C, the saturated vapour pressure at 20°C is 243 mm Hg; if it is active at the same thermodynamic activity, what is the concentration required to produce anaesthesia?

(From results of Suckling, *British Journal Anaesthesia* (1957), no. 29, p. 466.)

Homologous Series

The activities of *n*-aliphatic alcohols in a number of tests increase with chain length in a geometrical progression in a remarkably regular way. For example, when the logarithms of the concentrations required to produce comparable depressant effects in tadpoles are plotted against chain length, a straight line is obtained with maximum activity at *n*-undecanol; *n*-dodecanol is less active (Figure 5.1). The occurrence of this maximum is described as *cut off*. Similar straight lines are obtained when the logarithms of equiactive concentrations in tests on bacteria, inhibition of gut contractility, inhibition of oxygen consumption by lung, or ability to cause the release of histamine from lung, are plotted against chain length, though activity is not always seen to reach a maximum.

Once the aliphatic chain is longer than a few atoms, the standard free energy of formation of the compounds is increased by a regular increment for each additional methylene group. For homologous alkanes in the liquid state this is about 1 kcal (4.2 kJ)/mole/methylene group; for *n*-aliphatic alcohols it appears to be about 0.7 kcal (2.9 kJ)/mole/methylene group. Accordingly if μ_{01} is the standard chemical potential of one homologue and μ_{02} is that of the next homologue, the partial pressures, p_1 and p_2, at which the two have the same chemical potential, are given by the equation:

$$\mu_{01} + RT \ln p_1 = \mu_{02} + RT \ln p_2,$$

i.e.

$$\ln \frac{p_1}{p_2} = \frac{\mu_{02} - \mu_{01}}{RT} = \frac{a}{RT},$$

i.e. for solutions with the same chemical potential the log concentration will decrease linearly with chain length and for the *n*th member of the series

$$\ln p_n = \ln p_1 - \frac{na}{RT}.$$

For solutions of the same compound in two different phases the standard chemical potentials will be different in the two phases

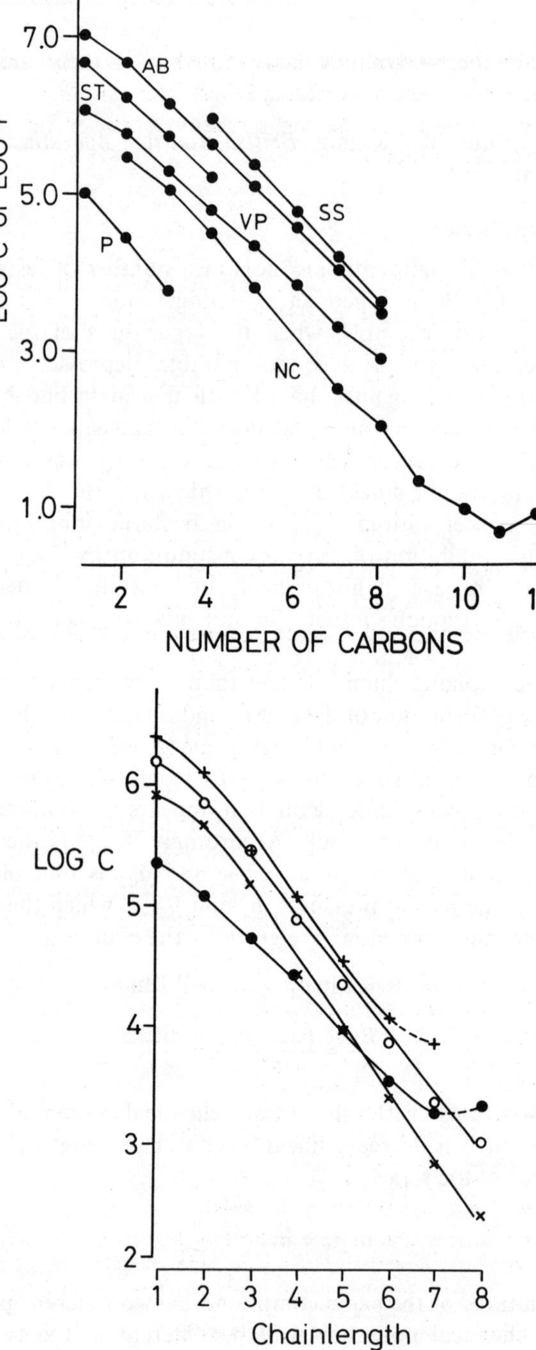

LOG C or LOG P
7.0
AB
ST
5.0
P
VP
SS
3.0
NC
1.0
2
4
6
8
10
12
NUMBER OF CARBONS
6
LOG C
5
4
3
2
1
2
3
4
5
6
7
8
Chainlength

(otherwise the compound will dissolve in both equally) and if C_1 is the concentration in the phase where the standard chemical potential is μ_0, and C_1' is the concentration in the phase where the standard chemical potential is μ_0',

$$\ln\left(\frac{C_1'}{C_1}\right) = \frac{\mu_0 - \mu_0'}{RT}.$$

If the standard chemical potential of the compound in one phase is increased by an increment a, so that μ_0 becomes $\mu_0 + a$ for the next homologue, and the same change occurs to the chemical potential in the second phase, the log partition coefficient,

$$\ln\left(\frac{C_2'}{C_2}\right) = \frac{\mu_0 - \mu_0'}{RT}$$

and is unaltered: this situation is unlikely. If the solvents are immiscible they must differ greatly in physical properties and the interactions between solute and solvent are likely to be much bigger for one solvent than for the other. If μ_0' becomes $\mu_0' + a'$ for the next homologue, and C_2 and C_2' are the concentrations with the same chemical potential as C_1 and C_1',

$$\ln\left(\frac{C_2'}{C_2}\right) = \frac{\mu_0 + a - \mu_0' - a'}{RT} = \ln\left(\frac{C_1'}{C_1}\right) + \frac{a - a'}{RT},$$

i.e. the log of the partition coefficient will be increased by a constant increment $(a - a')/2.3RT$, for the additional methylene group.

Ferguson considered that all the tests involved partitioning in some form. Even the effects on surface tension depend on

---

FIGURE 5.1 Properties of *n*-aliphatic alcohols.

(*A*) The number of carbon atoms is shown and the logarithm of the concentration (µM) which produces a saturated solution (SS), which inhibits the growth of *S. typhosus* (AB), which produces narcosis in tadpoles (NC), which lowers the surface tension of water to $63 \times 10^{-3}\,\mathrm{N\,m^{-1}}$ (ST), the logarithm of the vapour pressure (mm Hg $\times 10^4$ at 25°C: VP) and the logarithm of the partition coefficient between water and cotton seed oil ($\times 10^3$; P).
(From results of Meyer and Hemmi, *Biochim. Z.* (1935), no. 277, p. 39, and Ferguson, *Proc. Roy. Soc.*, B (1939), no. 127, p. 391.)

(*B*) The number of carbon atoms is shown and the logarithm of the concentration (µM) which depresses the contractility of guinea-pig intestine ($\times$), the motility of *paramecia* ($\bigcirc$), oxygen consumption by guinea-pig lung ($+$) and the release of histamine from guinea-pig lung ($\bullet$).
(From results of Rang, *British Journal Pharmacol.* (1960), no. 15, p. 185.)

differences between the molecules in the bulk of the solution and the molecules at the surface so the logarithms of the concentrations required to produce a particular depression of the surface tension of water, for instance, should decrease in a regular manner. The actual slope is about -0.5 ($=\log 3$) so the concentrations are in the ratio $1:\frac{1}{3}:(\frac{1}{3})^2:(\frac{1}{3})^3$, etc., a relationship which is known as Traube's rule.

The differences in slope can in some instances explain the occurrence of 'cut-off'. If the solubility decreases with chain length more than does the concentration needed to produce the particular effect, there will come a point at which it will no longer be possible to produce the effect because the homologue will not be soluble enough. The concentration of n-pentanol, for example, needed to inhibit the growth of *Staph. aureus* is the same as that of a saturated solution; higher homologues are inactive. It is not clear that this always accounts for cut-off. If the thermodynamic activity (p_t/p_s) is calculated it is observed that this is not strictly constant but tends to rise with increasing chain length. If it reaches 1, the vapour pressure of the solution producing the effect must be the same as that of a saturated solution, so insolubility should limit the activity of higher homologues, but there does not seem to be any explanation for the change in thermodynamic activity.

In homologous series it is to be expected that values of log partition coefficient will be linearly related to chain length but the slope of the line will vary according to the nature of the compounds and the solvent systems being studied. Bigger effects were observed, for instance, with partition between n-octanol and water than between *iso*-butanol and water.[3] It seems likely that the size of the effect depends primarily on the extent to which hydrogen bonds are broken in the transfer of the solute molecule from one phase to the other. As has already been mentioned (page 74) the partition coefficients for compounds in one pair of solvents can usually be satisfactorily related to partition coefficients for another pair of solvents.[4]

Results with homologous series confirm the idea that nonspecific biological activity can be related to physical properties. Though it is not known how such drugs produce their effects, it might be possible, nevertheless, to predict the activity of new compounds. For example, it is easy to make estimates of the activity of missing members in an homologous series. If this involves interpolation between members already tested the estimates should be very reliable. If the estimate involves extrapolation there is more uncertainty because of the possibility of cut-off.

Even when compounds are not members of a homologous series the prediction of nonspecific biological activity can be regarded as an extension of the prediction of partition coefficients or some similar property. The partition coefficient, P, of a solute in one system can be related to its partition coefficient, P', in another system by the expression,

$$\log P' = a \log P + b,$$

where a and b are constants (page 74). If P' is the partition coefficient between the biophase and water (or blood),

$$\log P' = \log C_{\text{bio}} - \log C$$

where C_{bio} is the effective concentration in the biophase and C is the measured concentration which is effective in the test. Accordingly,

$$-\log C = a \log P + b - \log C_{\text{bio}}.$$

If C_0 is the concentration of some standard compound which produces the same biological response and P_0 is its partition coefficient,

$$\log\left(\frac{C}{C_0}\right) = a \log\left(\frac{P}{P_0}\right),$$

i.e. log equipotent molar ratio $= a\pi$, if P and P_0 refer to n-octanol and water. If they refer to some other system, the constant, a, will have some other value. The relation between C and π can therefore be written,

$$\log C = a\pi + \log C_0,$$

i.e. log biologically effective concentration $= a\pi + a$ constant *or* $a \log P + a$ different constant.

EXAMPLE 5:3.

The concentrations of n-aliphatic alcohols which produced narcosis in tadpoles were: propanol, 0.11 M; butanol, 0.03 M; pentanol, 0.007 M; heptanol, 0.00038 M; octanol, 0.00013 M.

Calculate the concentration of n-hexanol which would be expected to produce this effect.

(From results of Meyer and Hemmi, *Biochem. Z.* (1935), no. 277, Table 11, p. 59.)

EXAMPLE 5:4.

The concentrations of saturated solutions of n-aliphatic alcohols at 25°C are: butanol, 1 M; pentanol, 0.25 M; hexanol, 0.063 M; heptanol, 0.016 M; octanol, 0.004 M. Calculate the expected solubilities of nonanol and decanol. The concentrations inhibiting the growth of *Staph. aureus* were: butanol, 0.8 M; pentanol, 0.25 M. Calculate the concentration of hexanol required to produce this effect. The concentrations inhibiting the growth of *S. typhosus* were: butanol, 0.45 M; pentanol, 0.13 M; hexanol, 0.039 M; heptanol, 0.012 M; octanol, 0.0034 M. Calculate the concentrations of nonanol and decanol which would be expected to produce this effect.

(From results of Meyer and Hemmi, *Biochem. Z.* (1935), no. 277, 127, Figure 2, p. 397.)

Biological Activity and Chemical Properties: the Ideas of Hansch

Where a drug produces its biological effects by interacting with some specific structure many chemical processes must be involved but these must ultimately all depend on the size and shape and on the distribution of electrons in the drug molecule. In theory this applies to all chemical reactions and as these can be correlated with considerable success with effects on ionisation assessed by Hammett's σ values, Hansch has suggested extending the application to the activity of drugs.[5]

The idea was originally applied to the biological activity of phenoxyacetic acids, which are selective weedkillers. They promote growth in a variety of plants which subsequently becomes excessive and they die. The compounds are selectively toxic to dicotyledons; monocotyledons, such as grasses and cereals are not usually killed. The activities of substituted phenoxyacetic acids can be estimated by comparing concentrations which produce a 10% growth of the test plants in standardised conditions in a standard time. In this situation the result would be influenced by the access of the drug across the cell wall, which will depend on its partition coefficient;

for series of compounds this can be expressed in terms of the π value for the substituent.

It is assumed that maximum activity in the conditions of the test depends upon the maximum probability of a molecule finding the active site and also on what it does to some key process when it has reached the site. If A is the probability of a molecule reaching a site of action in a given time interval, the rate of the biological response can be written

$$\frac{d(\text{response})}{dt} = ACk_x$$

where C is the external concentration and k_x the rate constant in the key process. It is assumed that there is some optimum value for the partition coefficient so the probability term A will depend on how close the partition coefficient of the compound is to this optimum. If the parameter π is used, rather than log partition coefficient, and if the distribution of values of π about π_0, the optimum value, is Gaussian, $A = a\,e^{-(\pi-\pi_0)^2/B}$ where a and B are constants and

$$\frac{d(\text{response})}{dt} = (a\,e^{-(\pi-\pi_0)^2/B})Ck_x.$$

In the tests on plants and on many other preparations, time forms part of the response and it is supposed that, when C represents the concentrations of compounds producing comparable biological responses, $d(\text{response})/dt$ is the same for all and is therefore a constant, r. Accordingly,

$$\log\frac{1}{C} = -\frac{\pi^2}{B} + \frac{2\pi\pi_0}{B} - \frac{\pi_0^2}{B} + \log k_x + \log(a/r).$$

If the rate constant, k_x, can be related to other rate constants by the Hammett relation ($\log k_x/k_H = \sigma\rho$, page 73), the equation can be rewritten,

$$\log\frac{1}{C} = a\pi^2 + b\pi + c\sigma + d$$

because π and σ are the only variables: a, b, c and d are a new set of constants.

When values of C have been obtained for a number of compounds for which the substituent constants π and σ are known, the method of least-squares can be used to obtain the values of a, b,

TABLE 5.2: *Activity of phenoxyacetic acids: C is the concentration producing* 10% *growth of the Avena* (*grass*) *samples in the standard time; P is the partition coefficient; π is the effect of the substituent on* $\log P$, *taking phenoxyacetic acid as standard, and σ is the effect of the substituent on the* pK_a *of benzoic acid.*

Substituent	σ	P	π	$\log 1/C$ (calc.)	$\log 1/C$ (obs.)
$3\text{-}CF_3$	0.55	320	1.09	6.8	6.5
$4\text{-}Cl$	0.37	168	0.80	6.3	6.4
$3\text{-}I$	0.28	325	1.08	6.1	6.3
$4\text{-}F$	0.34	43	0.20	5.0	6.3
$3\text{-}Br$	0.23	254	0.97	5.9	6.0
$3\text{-}SF_5$	0.68	1190	1.64	6.2	6.0
$3\text{-}Cl$	0.23	178	0.82	5.9	5.7
$3\text{-}NO_2$	0.78	29	0.04	5.7	5.3
$3\text{-}SCH_3$	-0.05	105	0.59	4.9	5.3
$3\text{-}C_2H_5$	-0.15	200	0.87	4.9	5.3
$3\text{-}n\text{-}C_3H_7$	-0.15	890	1.52	4.2	4.7
$3\text{-}OCH_3$	-0.27	36	0.13	3.1	4.7
$3\text{-}CN$	0.63	16	-0.23	4.1	4.5
$3\text{-}CH_3$	-0.17	75	0.44	4.3	4.3
$3\text{-}CH_3CO$	0.52	22	-0.08	4.5	4.0
$3\text{-}F$	0.06	41	0.18	4.2	3.5
H	0.00	27	0.00	3.4	3.5
$3\text{-}OH$	-0.36	5	-0.73	-1.8	3.7
$3\text{-}COOH$	0.27	21	-0.13	3.5	3.0
$3\text{-}n\text{-}C_4H_9$	-0.15	3100	2.03	2.4	0.0

(From results of Hansch, Maloney, Fujita and Muir, *Nature* (1962), no. 194, p. 178.)

c and d which give the best fit (see Appendix). There are now tables of values of σ for most substituents and also for many values of π (Table 2.4), though these can always be measured as described in Chapter 2. In theory, the values of a, b, c and d could then be used to predict the value of $\log 1/C$ for any new compound for which the values of π and σ are known. It should be noted that if the coefficients a and b are of opposite sign they will reproduce the effects of cut-off (though the maxima in Figure 5.1 are too sharp for a parabola and suggest a higher power of π). This is in fact implicit in Hansch's assumption that there is some optimum value for π.

The goodness of fit can be assessed from the *correlation coefficient*, r, which should be close to 1, and the residual variance, which should be small (see Appendix).

Results for 20 phenoxyacetic acids are shown in Table 5.2. The calculated values were obtained with the equation:

$$\log \frac{1}{C} = 4.08\pi - 2.14\pi^2 + 2.78\sigma + 3.36.$$

For the correlation of the observed and calculated values of $\log(1/C)$, $r = 0.66$. In general, the difference between the observed and calculated values of $\log(1/C)$ is about 0.3 log units, indicating a discrepancy of only a factor of 2, but with the 4-fluoro and 3-methoxy compounds the discrepancy is by a factor of 20 and with the 3-hydroxy and 3-*n*-butyl compounds the 'prediction' is a complete failure. There is some uncertainty about some of the values of π and σ and in a further study[6] the equation for the least-squares fit with 34 compounds of this type was calculated to be

$$\log \frac{1}{C} = 3.24\pi - 1.97\pi^2 + 1.86\sigma + 4.16,$$

with $r = 0.78$.

It was also observed that the activity of 11 analogues of chloramphenicol against the Gram positive organism, *Staph. aureus*, could be expressed by the equation:

$$\log A = 0.48\pi - 0.54\pi^2 + 2.14\sigma + 0.22,$$

with $r = 0.95$ and for activity against the Gram negative organism, *E. coli*, the equation was:

$$\log A = 0.36\pi + -0.74\pi^2 + 1.82\sigma + 0.62,$$

with $r = 0.82$.

For 35 phenols[7] the log of the activity expressed relative to phenol itself (i.e. the log of the 'phenol coefficient') against *Micrococcus pyogenes* depended primarily on the partition coefficient; for the equation:

$$\log A = 0.951\pi + 0.144,$$

had $r = 0.98$ and the inclusion of terms for π^2 and σ gave the equation

$$\log A = 0.953\pi - 0.001\pi^2 + -0.201\sigma + 0.134$$

also with $r = 0.98$. For activity against *S. typhosus* the equations were:

$$\log A = 1.300\pi - 0.283\pi^2 - 0.328\sigma + 0.129, \quad r = 0.92;$$
$$\log A = 0.454\pi + 0.477, \quad r = 0.73,$$

and
$$\log A = 1.312\pi - 0.288\pi^2 + 0.139; \quad r = 0.92.$$

With this organism, therefore, there is a need for both π and π^2, and evidence of 'cut-off', although the contribution from σ appears small.

The relative importance of π and σ would be expected to depend on the type of test, as well as on the type of compound. When the test is such that activity depends primarily on rates of access, π and π^2 should be important and σ should only matter if the compounds are ionisable and it determines the proportion able to enter cells. For example, results with 28 compounds tested as general depressants on tadpoles could be related[8] to partition coefficients by the equation:

$$\log\frac{1}{C} = 0.869\log P + 1.242,$$

with $r = 0.96$. A high correlation with physical properties would be expected anyway in this test, however, because the compounds are acting nonspecifically (compare this equation with the answer to Example 5:1). Values of π were likewise dominant in tests of nonspecific activity on the frog heart, on the motility of paramecia, on the contractility of guinea-pig intestine[9] and on the effects of substituted benzoic acids on mosquito larvae,[7] as well as on the actions of phenols on *Micrococcus pyogenes*, already referred to.

Both π^2 and π terms were found to be important in tests on the narcotic effects of barbiturates and their effects on oxygen uptake by slices of rat brain,[9] on the carcinogenic activity of substituted azobenzenes[7] as well as on the actions of phenols on *S. typhosus*, already referred to. The relative importance of the π^2 term depends on the test.[10]

In tests on the toxic effects of substituted phenethylphosphates on house-flies the σ term was most important.[7] In tests on the local anaesthetic properties of esters of diethylaminoethanol with substituted benzoic acids, and on the inhibition of cell division in *arbacia* eggs by barbiturates, both π and σ terms were important but the coefficient for π^2 was small.[7] In very many other tests, however, all the coefficients appeared to be important.

The equation can easily be modified to include other parameters which are related to chemical properties (see ref. 11). These include the Taft polar constant, σ^*, or an inductive constant, σ_i, for effects in aliphatic amines. Group dipole moments, μ, have also been used. Values of ΔR_m have been used instead of π. To try to allow for the size of groups the Taft steric parameter, E_s, has been included. This is derived from the observed effects of groups in chemical reactions where these are thought to be attributable only to the size of the substituent. Some workers have used molal volumes, V_m, for this purpose.

There is therefore a large range of parameters expressing chemical properties which have been correlated with $\log 1/C$; these are mostly available in tables.[11] With so many to choose from and usually four coefficients, a, b, c and d, to be solved, it should not be difficult to obtain a correlation coefficient better than 0.9, if the number of compounds tested is small. It is doubtful whether a correlation based on results with less than about 20 compounds is of much value.

A quantitative correlation between biological activity and chemical properties has two possible applications. It might be used predictively in the design of new compounds and it might be used to try to deduce the types of chemical process involved in biological activity.

If results have been obtained with a trial group of compounds in which the substituents have a wide range of values of π and σ, or whatever other parameters are being used, it should be possible to see from the fitted equation what these values should be for compounds with high activity. Substituents with these values could then be found from tables of π and σ, so it should be possible to concentrate on making compounds likely to be active and avoid making and testing a lot of weak or inactive drugs.

Predictions made in this way are much more likely to be reliable if they involve interpolation within the range of π and σ already studied, rather than extrapolation outside this range. A good correlation is obtained, for instance, within a range of n-aliphatic alcohols (see Examples 5:3 and 5:4) but extrapolation outside the range fails to predict cut-off. Even with a good correlation coefficient, predictions involving extrapolation outside *spanned substituent space* are uncertain.[12]

Predictions involving values of parameters, such as π or σ, within the range already studied should be much more reliable and may make it possible to find new highly active compounds, if the particular combination of parameters needed for this has not already been tested. It is, however, possible to have a good overall correlation with one or two compounds which are aberrant. It may not matter much if a compound turns out to be less active than predicted but it does matter if a compound is predicted to be inactive when it is really active. For example, 4-fluorophenoxy-acetic acid is 20 times as active as its calculated value in the original study (Table 5.2). Even if there is only a slight chance, say 1 in 100, of missing an active compound because it is predicted to be inactive, many medicinal chemists would find this risk unacceptable.

It is possible to argue that these aberrant compounds do not really belong to the group but to be of any use to the chemist, the definition of a group must be made in terms of recognisable structural features. The difference often found between optical enantiomers, for instance, is a major problem but although physical parameters such as π and σ are the same for both enantiomers, there is a recognisable chemical basis for treating them as separate groups. In many pharmacological tests, however, the enantiomers are known to be acting at the same receptors and there is no pharmacological reason for treating them separately.

The variable quality of the biological results is another major problem. In many tests it is not clear whether the compounds produce effects with comparable time-courses and have parallel log dose response curves and, indeed, if it is really legitimate to attempt a quantitative comparison between them (page 30). The term 'biological activity' may cover types of result which can be measured with an accuracy of $\pm 10\%$, the most accurate likely to be obtainable, or types of result where the accuracy is at best a factor of 2, i.e. results lie in the range $r/2$ to $2r$, where r is the mean response. In addition, no allowance is made for differences in variance within results for a group of compounds. It is not unusual to find that estimates for very active or very inactive compounds are less accurate than those for intermediate compounds and this may bias the calculation of the regression equation. It has, nevertheless, been claimed that within spanned substituent space 'you can expect...to be able to make a prediction close to your standard error of your equation.'[12]

The predictive value of equations of the Hansch type, even within spanned substituent space, is arguable, however, and this applies to other attempts at quantitative structure-activity relations (QSAR). It is also true of attempts which have been made to deduce from such equations how the drugs may be interacting with receptors. For instance, the high importance of π has been taken to indicate hydrophobic bonding between drug and receptor[10] but when considering possible actions at a molecular level it is at least as important, if not more important, to examine the nature of the pharmacological test as it is to examine the equation. Hansch originally assumed that the size of the response depends on the rate of arrival of drug at the active site and that there is some optimum partition coefficient for drug access (page 197). In this situation activity depends primarily on transport. In other tests transport may be less important but activity still depends on many factors and cannot be taken to depend simply on binding. With compounds which are agonists, for instance, activity clearly depends on efficacy as well as on affinity.

Tests where activity depends mainly on transport yield little information about real biological activity, i.e. effects at receptors. If correlations are to be applicable to biological activity in general, as well as to transport processes, it is necessary to consider how far activity at receptors is predictable. To investigate this it seems reasonable to begin by inquiring how far affinity is predictable.

Is Affinity Predictable?

The most fundamental property of a drug is its affinity for the site where it acts and if it is an antagonist, tested in a situation where there is no complication from problems of transport, this is the only factor which determines biological activity. If drug activity is to be predictable in more complex tests, such as many of those whose results have been fitted to equations of the Hansch type, it ought first to be possible to predict the affinity of antagonists.

One of the simplest changes in structure which can be made is to replace a methyl group on a quaternary nitrogen atom by an ethyl group. The increase in size might provide the opportunity for extra binding between drug and receptor and the affinity should increase. If the Gibbs free energy for the adsorption of the original compound to the receptor is $-\Delta G$, that for the homologue could be $-(\Delta G + a)$, where a is the increment due to the extra methylene group. If the affinity constant is measured in conditions in which the antagonists have had time to come into equilibrium with the receptors,

$$\ln K = -\frac{\Delta G}{RT}$$

for the original compound and

$$\ln K' = -\frac{(\Delta G + a)}{RT}$$

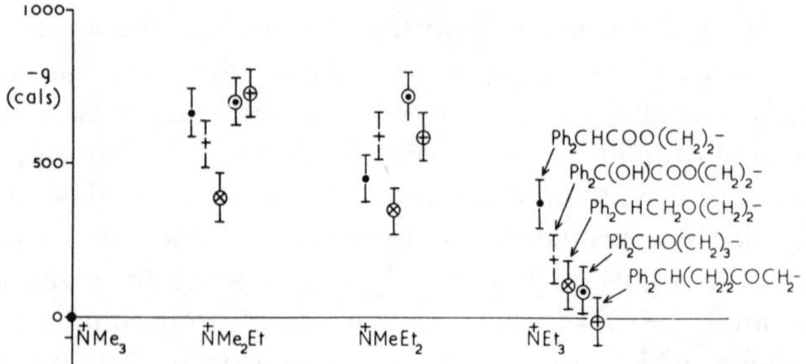

FIGURE 5.2 The effect of altering the onium group on the adsorption of some antagonists of acetylcholine at the muscarinic receptors of the guinea-pig ileum. The change in log affinity constant·has been converted into the corresponding change in free energy, $-g$ $(= \Delta\Delta G)$, shown in $\mathrm{cal\,mol^{-1}}$. Values for the five series of compounds have been separated laterally for identification. Note that the effects are similar in all the series.
(From results of Barlow, Scott and Stephenson, *British Journal Pharmacol.* (1963), no. 21, p. 509.)

for the homologue so,

$$\ln\frac{K'}{K} = \frac{-a}{RT}.$$

The affinity should therefore increase (a is negative) and the ratio of K'/K should be independent of the affinity of the original compound, i.e. similar changes should be seen in affinity when the same chemical change is made in different compounds, provided they are interacting with the same receptor.

These ideas can be tested. The affinity constants of antagonists of acetylcholine for 'muscarine-sensitive' acetylcholine receptors can be measured on the isolated guinea-pig ileum in conditions in which they have had time to come into equilibrium with the receptors (page 32). The error in mean estimates of $\log K$ is almost always less than $0.1 \log$ unit. The effects of successively replacing methyl groups by ethyl in five series with the general structure:

$$\mathrm{R\overset{+}{N}Me_3, \quad R\overset{+}{N}Me_2Et, \quad R\overset{+}{N}MeEt_2, \quad R\overset{+}{N}Et_3}$$

are shown in Figure 5.2. Although the affinity constants of the trimethylammonium compounds differed up to 200-fold, the changes in affinity produced by replacing methyl groups by ethyl are very similar in all five series.[13] Unfortunately, when further compounds were tested and the changes in the onium group included replacing methyl groups by pyrrolidine and piperidine

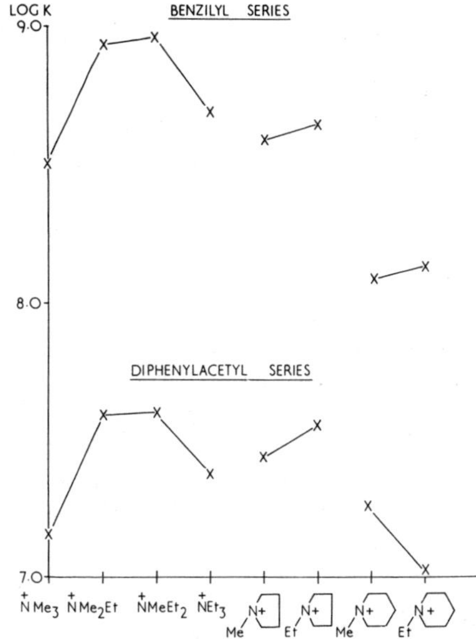

FIGURE 5.3 The effects of extending two of the series in Figure 5.2. Values of log affinity constant are shown and the striking parallel observed with the replacement of methyl by ethyl in the onium group does not extend to the compounds in which the onium group is part of a pyrrolidine or piperidine ring.
(From results of F. B. Abramson, PhD thesis (1964), University of Edinburgh.)

rings there were definite differences between series (Figure 5.3). The pyrrolidyl and piperidyl groups lower the affinity of the benzilic esters much more than they lower the affinity of the diphenylacetic esters.

The Variable Effects of Groups on Affinity

The finding that the effects of groups on affinity are not constant is extremely important. Even in chemical processes there are usually some points which do not fit a Hammett relation, where a particular combination of groups does not produce the expected effects. If this is a more common situation with affinity for receptors, the biological activity cannot be regarded as being made up of additive components from each part of the molecule, as was suggested by Free and Wilson[14] in an empirical approach to the quantitative analysis of structure-activity relationships. The contribution made by one group will be affected by the presence of others.

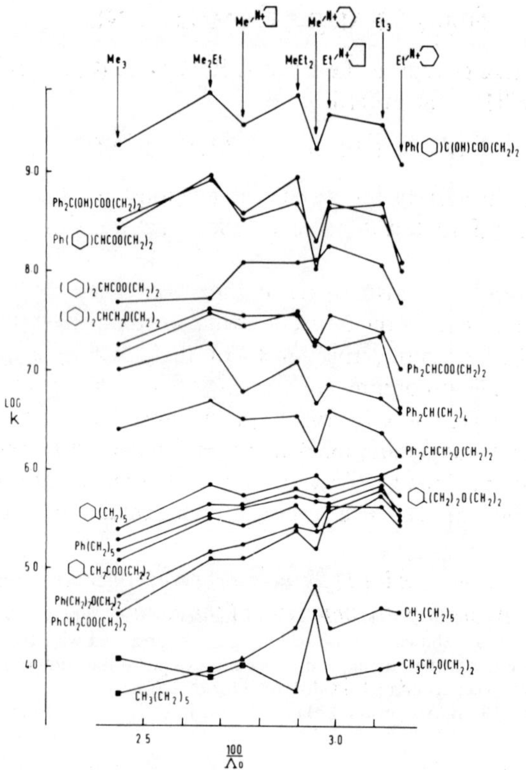

FIGURE 5.4 Values of $\log K$ for series of compounds acting at the muscarinic acetylcholine receptors of the guinea-pig ileum. The onium groups are arranged in increasing order of size, estimated from measurements of electrical conductance ($100/\Lambda_0$ correlates excellently with increments in apparent molal volume, ϕ_v^0). Note that the effects of the onium groups on affinity differ from one series to another but in series with two large groups at the far end there is some similarity (a saw-tooth pattern) which is different from those with only one large group (an uphill pattern). (From results of Abramson, Barlow, Mustafa and Stephenson, *British Journal Pharmacol.* (1969), no. 37, p. 223.)

The extent to which effects of groups on affinity may vary can be assessed from results obtained with series of compounds having eight onium groups at one end chef the molecule and phenyl and/or cyclohexyl groups at the other;[15] these are shown in Figure 5.4. Changes in the onium group produce different effects in different series but there are some trends within particular types of series. When the group at the far end is CH_3—, $PhCH_2$—, or $C_6H_{11}CH_2$—, the affinity increases in the order

$$\overset{+}{N}Me_3 < \overset{+}{N}Me_2Et < \overset{+}{N}MeEt_2 < \overset{+}{N}Et_3.$$

With two large groups at the far end, Ph_2CH—, $(C_6H_{11})_2CH$—, or $Ph(C_6H_{11})CH$—, the order is:

$$\overset{+}{N}Me_3 < \overset{+}{N}Me_2Et \doteqdot \overset{+}{N}MeEt_2 > \overset{+}{N}Et_3.$$

The change in affinity for a particular change in structure is not constant, therefore, but depends on the structure of the rest of the molecule.

This can be seen also by examining the effects on affinity of altering the far end of the molecule while keeping the onium group constant. For example, the effects of replacing hydrogen in *n*-pentyltrimethylammonium by phenyl can be calculated by subtracting $\log K$ for *n*-pentyltrimethylammonium from $\log K$ for phenylpentyltrimethylammonium; seven more estimates of this effect can be made from the values for the other members of these series (Table 5.3*A*). The contributions made by phenyl or

TABLE 5.3*A*: *Values of* $\log K$ *for muscarinic acetylcholine receptors of the guinea-pig ileum for compounds of the series:*

$$X—CH_2CH_2CH_2CH_2\overset{+}{N}\diagup_{\diagdown}$$

	X = CH$_3$—	PhCH$_2$—	Δ	Ph$_2$CH—	Δ	$\Delta\phi_v^0$ cm^3 mol^{-1}
$\overset{+}{N}Me_3$	3.73*	5.18	1.45	7.01	1.83	0
$\overset{+}{N}Me_2Et$	3.97*	5.55	1.58	7.27	1.72	15
Methylpyrrolidinium	4.16	5.60	1.44	6.79	1.19	19
$\overset{+}{N}MeEt_2$	4.40	5.73	1.33	7.09	1.36	30
Methylpiperidinium	4.81	5.69	0.88	6.66	0.97	33.4
Ethylpyrrolidinium	4.37	5.65	1.28	6.86	1.21	34.4
$\overset{+}{N}Et_3$	4.59	5.85	1.26	6.71	0.86	44.5
Ethylpiperidinium	4.55	5.47	0.92	6.58	1.11	48
		mean	1.27	mean	1.28	
		s.d.	0.25	s.d.	0.34	

Δ indicates the difference in $\log K$ due to the presence of a phenyl group; $\Delta\phi_v^0$ is an estimate of the increase in size, assessed from the apparent molal volume at infinite dilution (page 81), produced by altering the onium group. An asterisk indicates that the compound is an agonist and the affinity was measured using an irreversible blocking agent; underline indicates that the compound is a partial agonist.

TABLE 5.3*B*: *Average effects* ($\Delta \log K$) *of replacing hydrogen by phenyl or hydrogen by* cyclohexyl *in the series.*

	H → Ph	$\Delta \log K$	H → C_6H_{11}	$\Delta \log K$
n-pentyl		1.3		1.5
Ethoxyethyl				
(EtOCH$_2$CH$_2$—)		1.3		1.7
Phenylpentyl				
(PhCH$_2$CH$_2$CH$_2$CH$_2$CH$_2$—)		1.3		
Phenylethoxyethyl				
(PhCH$_2$CH$_2$OCH$_2$CH$_2$—)		1.1		1.6
Phenylacetoxyethyl				
(PhCH$_2$COOCH$_2$CH$_2$—)		2.1		3.3
*Cyclo*hexylacetoxyethyl				
(C$_6$H$_{11}$CH$_2$COOCH$_2$CH$_2$—)		3.0		2.5

(From results of Abramson, Barlow, Mustafa and Stephenson, *British Journal Pharmacol.* (1969), no. 37, p. 207.

cyclohexyl groups in several different series can be examined in this way and the mean values are shown in Table 5.3*B*. The standard deviation of the mean was usually 0.3–0.4.

In the first four sets of comparisons, involving 32 pairs of compounds, the effects of introducing the phenyl group are reasonably consistent, but in the fifth set the phenyl group makes a bigger contribution to binding and in the last set the contribution is very much bigger indeed. Similarly the replacement of hydrogen by cyclohexyl produces much the same average effects in three of the series but a much bigger effect in the cyclohexylacetoxyethyl series and a very much bigger effect indeed in the phenyl-acetoxyethyl series.

When the introduction of a substituent which could potentially lead to increased affinity fails to do so, it can be argued that it is not in a position to make its full contribution. Possibly the part of the molecule to which it is attached has become too big to fit the space available. Possibly it impedes rotation within the molecule and so sterically hinders groups which are important for binding from assuming a suitable conformation. It is not difficult to think of reasons why a group fails to bind as well as might be expected.

It is much more difficult to explain why a group binds very much more strongly than expected, as occurs with these compounds. It is possible, but unlikely in these flexible molecules, that the substituent alters the preferred conformation so that the molecule is now held in a shape particularly favourable for binding. It is also

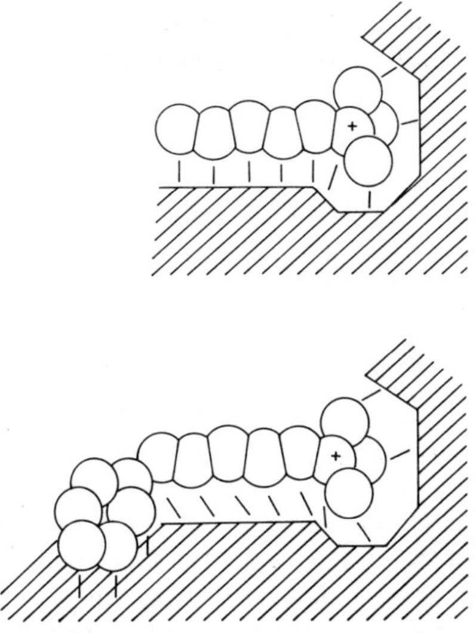

FIGURE 5.5 A scheme to illustrate the disturbance to existing binding caused by introducing a substituent which also contributes to binding. The upper picture represents the starting molecule, binding as well as possible. The lower picture shows that the interaction between the new group and the receptor involves displacement of the original fragment and weakens its contribution to binding; the new group, likewise, cannot exert its full effect.

possible that there is a form of apparent cooperativity, with the binding of one group making it easier for another group to be bound more strongly, but this implies that in the smaller molecules these groups are not binding as well as they could do. It seems likely that with any drug containing a number of groups which can contribute to binding, these will not all be able to make their full contribution. If a phenyl group is introduced into a small molecule such as *n*-pentyltrimethylammonium, it can make a big contribution to binding but this will probably be offset to some extent by a disturbance in the existing binding of the pentyl part (Figure 5.5).

If the disturbance produced by introducing a phenyl group is similar to that produced by introducing a cyclohexyl group, it may be possible to introduce a cyclohexyl group into a compound with a phenyl group without the need for much rearrangement at the receptor so the cyclohexyl group will make a bigger contribution to binding (because it is not accompanied by much further disturbance). From the results in Table 5.3*B* it is possible to obtain some idea of the size of the group contributions to binding and the disturbance terms. The changes in $\log K$ indicate an average contribution to binding of $1.8\,\text{kcal}\,\text{mol}^{-1}$ ($7.5\,\text{kJ}\,\text{mol}^{-1}$) for a phenyl group and $2.2\,\text{kcal}\,\text{mol}^{-1}$ ($9.2\,\text{kJ}\,\text{mol}^{-1}$) for cyclohexyl. For the two together the contribution is $6.5\,\text{kcal}\,\text{mol}^{-1}$ ($27.2\,\text{kJ}\,\text{mol}^{-1}$) and if this involves only one disturbance, not two, the loss caused by disturbance is $2.5\,\text{kcal}\,\text{mol}^{-1}$ ($10.5\,\text{kJ}\,\text{mol}^{-1}$). Thus the phenyl group could contribute $4.3\,\text{kcal}\,\text{mol}^{-1}$ ($18.0\,\text{kJ}\,\text{mol}^{-1}$) but this is reduced to $1.8\,\text{kcal}\,\text{mol}^{-1}$ ($7.5\,\text{kJ}\,\text{mol}^{-1}$) and likewise the cyclohexyl group could contribute $4.7\,\text{kcal}\,\text{mol}^{-1}$ ($19.6\,\text{kJ}\,\text{mol}^{-1}$) but this is reduced to $2.2\,\text{kcal}\,\text{mol}^{-1}$ ($9.2\,\text{kJ}\,\text{mol}^{-1}$).

This explanation assumes a static receptor, which is consistent with the compounds all being competitive antagonists. There is no need to suppose that there is cooperativity between groups involving conformational changes in the (antagonist) receptor. If the ideas are correct it should be possible to obtain some idea of the 'true' contribution made by a group, in a situation where its introduction causes no disturbance, by observing its maximum effect in as many compounds as possible. Measurements have been made of the affinities of altogether about 200 compounds for the muscarinic acetylcholine receptor of the guinea-pig ileum.[16] The contribution made by a phenyl group can be seen as a histogram in Figure 5.6; this shows the number of instances, n, in which the

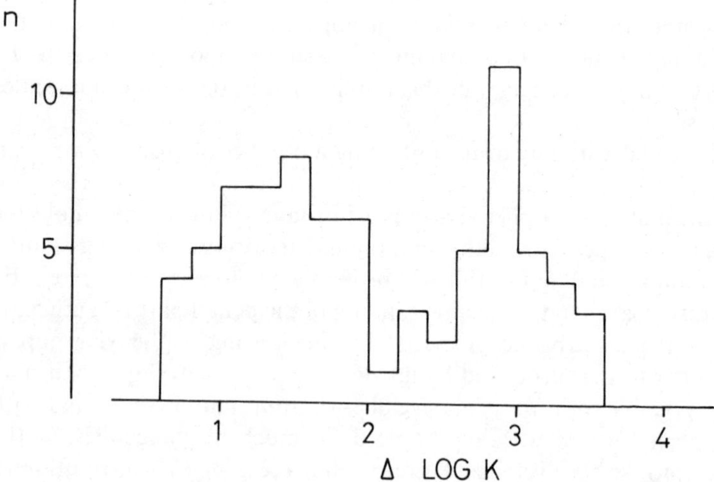

FIGURE 5.6 The effects on affinity for muscarinic acetylcholine receptors of the guinea-pig ileum of replacing hydrogen by phenyl in 77 compounds. The histogram shows the number of instances (n) in which the change in $\log K$ ($\Delta\log K$) lies within the range indicated.

TABLE 5.4: *Effects of changes in structure on* $\log K$ *for the muscarine-sensitive acetylcholine receptors of the guinea-pig ileum and on size, estimated from* ϕ_v^0.

The extreme values of the change in $\log K$ *($\Delta \log K$) are shown together with the range, number of comparisons and an estimate of the average change in apparent molal volume at infinite dilution ($\Delta \phi_v^0$).*

	$\Delta \log K$				$\Delta \phi_v^0$
	max.	min.	range	n	
Replacement of $-\overset{+}{N}Me_3$ by:					
$\overset{+}{N}Me_2Et$	0.722	-0.187	0.909	23	15.6
$\overset{+}{N}MeEt_2$	0.893	-0.339	1.232	23	30.0
$\overset{+}{N}Et_3$	1.252	-0.303	1.555	23	44.7
methylpyrrolidinium	0.614	-0.227	0.841	23	19.9
ethylpyrrolidinium	1.035	-0.191	1.226	23	34.6
methylpiperidinium	1.082	-0.477	1.559	23	33.6
ethylpiperidinium	0.992	-0.654	1.646	23	48.0
Replacement of $-H$ by:					
Ph	3.507	0.624	2.883	77	63
C_6H_{11}	4.041	0.875	3.166	60	79
OH	1.938	-1.492	3.430	59	1
Replacement of $-CH_2CH_2-$ by:					
$-CO-O-$	0.700	-0.647	1.347	24	-13.8
$-CH_2-O-$	0.341	-0.664	1.005	32	-10.9

(From results of Abramson, Barlow, Franks and Pearson, *British Journal Pharmacol.* (1974), no. 37, p. 207.)

change in $\log K$ falls within the range indicated. In three instances the change is between 3.4 and 3.6 log units but there is a very wide range of effect; in four instances the increase was only between 0.6 and 0.8 log units. Results for other changes in structure are shown in Table 5.4 and there is a similar wide variation with other groups.

If the maximum effects seen in Table 5.4 represent the contribution which a group can make to binding when its introduction causes no disturbance to existing binding, they should be related to the chemical properties of the group. For nonpolar groups, i.e. those containing only carbon and hydrogen, the maximum effect appears to be related to the size of the substituent in solution as indicated by the increment in apparent molal volume at infinite dilution, $\Delta \phi_v^0$ (Figure 5.7A). For the replacement of methyl by ethyl the extra methylene group seems capable of contributing about $1 \, kcal \, mol^{-1}$ ($4.2 \, kJ \, mol^{-1}$) to the free energy of

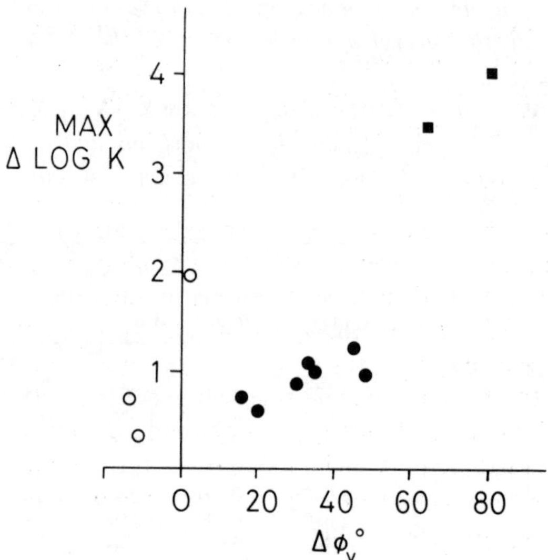

FIGURE 5.7(*A*) The maximum contributions of groups to affinity plotted against the change in size. For alkyl groups (●) and phenyl and cyclohexyl groups (■) it appears that the contribution is related to the size but this is not true for changes involving oxygen (○).

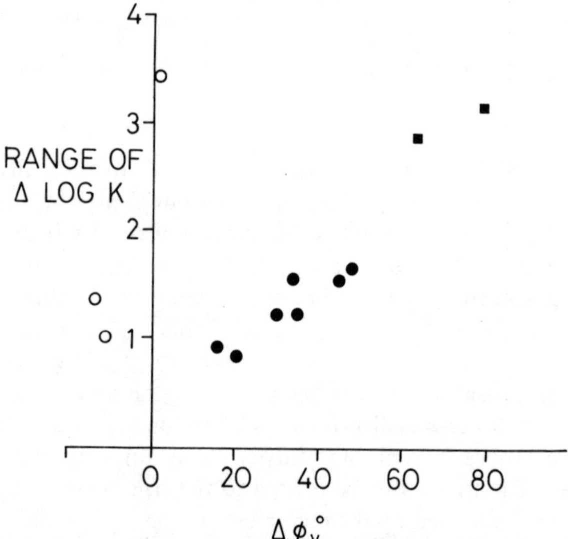

(*B*) The range of contribution plotted against the change in size. The range appears to be related to size for groups containing only carbon and hydrogen, but the large variation in the effects of the hydroxyl group is not related to its size, which is very small in water.

adsorption ($\Delta \log K = 0.7$). This is much higher than would be expected from van der Waals' forces and suggests that the binding is largely hydrophobic. For the cyclohexyl group, with 6 methylene groups, the maximum contribution observed is $5.7\,\text{kcal}\,\text{mol}^{-1}$ ($23.8\,\text{kJ}\,\text{mol}^{-1}$) and for the phenyl group, which appears to be rather smaller than the cyclohexyl group in water, the maximum contribution observed was $5.0\,\text{kcal}\,\text{mol}^{-1}$ ($20.9\,\text{kJ}\,\text{mol}^{-1}$). The larger onium groups never produced the binding which might be expected from their number of methylene groups, possibly because there is limited space in that part of the receptor. When alkyl groups on the onium nitrogen were incorporated in a ring the maximum effects were smaller still (compare triethylammonium, ethylpyrrolidinium and methylpiperidinium in Table 5.4) which again suggests that space is limited and the relative inflexibility of the rings means that they invariably disturb the binding of the rest of the molecule to some extent.

The effects of a hydroxyl group, however, do not appear to be

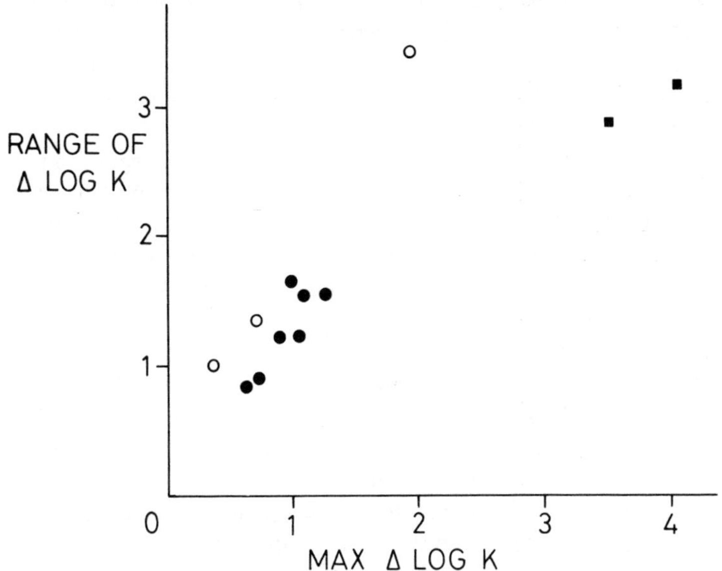

(*C*) The range plotted against the maximum effect. Whatever the group, big maximum effects are associated with a big range, so attempts to produce a big increase in affinity have a big chance of falling short of what is hoped for.
(From results of Abramson, Barlow, Franks and Pearson, *British Journal Pharmacol.* (1974), no. 37, p. 90.)

related to its apparent size, which in water is very small indeed. The biggest effect on $\log K$ was an increase of 1.9 log units, corresponding to a 100-fold increase in affinity and to a contribution of $2.7\,\text{kcal}\,\text{mol}^{-1}$ ($11.3\,\text{kJ}\,\text{mol}^{-1}$) to the free energy of adsorption. This is weak for a hydrogen bond (page 183) but in keeping with the idea that, in water, which is itself strongly hydrogen-bonded, such bonds will not be as strong as might otherwise be expected. Other changes involving oxygen atoms, replacement of $-CH_2CH_2-$ by $-CH_2O-$ and $-CO-O-$, produce smaller maximum effects. With these the possible gain in affinity due to coulombic forces, between the receptor and the polar oxygen atom and/or polarised carbonyl group will be offset by some loss in hydrophobic bonding.

If the maximum effects depend on the chemical properties of groups, such as size and polarity, they should be predictable but as can be seen, the actual effects of a group are very variable. For groups containing only carbon and hydrogen the range appears to depend on the size (Figure 5.7B); with large groups there is a large variation. For hydroxyl groups the range is unrelated to size but for all the groups studied the range appears to be related to the maximum effect (Figure 5.7C). Groups which could make the greatest contribution to binding are also groups where the variation in the effect is greatest.

It is reasonable to expect that the affinity should depend on the sum of all the forces between the drug and the receptor. These may be divided into components from various groups and if these are additive it should be then possible to predict values for new combinations of groups. It is clear, however, that the contribution made by one group is affected by the presence of another group. This in fact occurs with chemical properties when groups can interact with each other as, for example, when they are in the *o*-position. It occurs to a very much greater extent, however, with drugs interacting with a complex receptor. Changes in the composition of the onium group at one end of a molecule, for example, can disturb the binding of groups at the far end, leading to changes in stereospecificity.[17] The problem in predicting affinity is to know the extent to which it will fall short of the maximum possible and from Figure 5.7C there is the warning that with groups which could produce a big effect there is also a big variation. It is biological information which is needed to solve this problem, which involves binding to a highly structured receptor.

Chemical parameters by themselves do not give this information and affinity is not predictable until more is known about receptor structure.

At present the most that appears possible is to set an upper limit to affinity. For example, starting with a value of 3.7 for $\log K$ for *n*-pentyltrimethylammonium, the maximum effect to be expected by changing $-CH_2CH_2-$ to $-CO-O-$ and replacing hydrogen by phenyl would be to increase $\log K$ by $0.7+3.5$ log units (Table 5.4) so the upper limit to the value for phenylacetylcholine would be 7.9. The experimental value is 4.5.

EXAMPLE 5:5.

For the muscarine-sensitive acetylcholine receptors of the guinea-pig ileum the value of $\log K$ for *n*-pent$\overset{+}{N}Me_3$ is 3.7. Use the maximum and minimum increments listed in Table 5.4 to suggest limiting values of $\log K$ for phenylcyclohexylglycolloyl-choline ($Ph(C_6H_{11})C(OH)COOCH_2CH_2\overset{+}{N}Me_3$), i.e. the values it should not be less than or greater than. The value of ϕ_v^0 for pentyltrimethylammonium iodide is $189.7\,cm^3\,mol^{-1}$; calculate the value for phenylcyclohexylglycolloylcholine. The experimental values are $\log K = 9.65$ (for the R-isomer) and $\phi_v^0 = 318.7$.

Entropy and the Importance of Water

The binding of antagonists to receptors in conditions of equilibrium (in so far as this can be obtained with living cells) is the simplest process which can be studied in molecular pharmacology. The variable effects which a substituent, such as a phenyl group, has on the binding of closely related compounds suggests that it has different effects on the entropy of adsorption in different compounds, so results cannot be expected to fit Hammett or Hansch equations, which depend upon free energy increments being predictable (page 77).

It should be possible to assess the effects of groups on the entropy of adsorption from estimates of the enthalpy of adsorption (ΔH), obtained from measurements of affinity at different temperatures (page 44). Belleau, Tani and Lie,[18] working with alkyltrimethylammonium salts and acetylcholinesterase from ox red cells, measured inhibitor constants at 10°, 20°, 25° and 30°C and obtained the results shown in Table 5.5. The inhibitory activity

TABLE 5.5: *Free energy* (ΔG), *enthalpy,* (ΔH) *and entropy* (ΔS) *of binding of alkyltrimethylammonium salts with acetylcholinesterase from ox red cells. The compounds are reported to behave as competitive antagonists (but see page 149) and the inhibitor constant,* K_i *was measured at* $10°, 20°, 25°$ *and* $30°C$ *and the results used to calculate* ΔH*. Only values of* K_i *at* $25°C$ *are shown;* K *is the affinity constant* $(1/K_i)$*.*

Chain length (number of carbon atoms)	K_i (25°C)	$\log K$	$-\Delta G$ $kcal\,mol^{-1}$	$-\Delta H$ $kcal\,mol^{-1}$	ΔS $cal\,mol^{-1}\,deg^{-1}$
1	2.33 mM	2.63	3.59	6.60	-10.1
2	1.60	2.80	3.81	6.45	-8.9
3	1.34	2.87	3.92	6.32	-8.1
4	0.84	3.08	4.20	5.22	-3.4
5	1.74	2.76	3.76	5.40	-5.5
6	1.34	2.87	3.92	4.55	-2.1
7	1.02	2.99	4.08	4.49	-1.4
8	0.66	3.18	4.34	4.40	-0.2
9	0.48	3.32	4.53	4.40	$+0.44$
10	0.226	3.65	4.97	4.26	$+2.4$
11	0.116	3.94	5.37	2.76	$+9.1$
12	0.0052	4.28	5.85	2.75	$+10.4$

(From results of Belleau, Tani and Lie, *Journal American Chem. Soc.* (1965), no. 87, p. 2285; Belleau, *Ann. NY Acad. Sci.* (1967), no. 144, p. 708.)

increases in a regular manner from $Me_4\overset{+}{N}$ to n-Bu$\overset{+}{N}Me_3$, declines with n-pent$\overset{+}{N}Me_3$, and then rises again regularly. The entropies of adsorption, however, show an almost continuous increase with chain length, indicating increased disorder. In general, the adsorption is exothermic (ΔH is negative) and affinity is greater at lower temperatures, but in a study of further compounds[19] some were found to have lower affinity at lower temperatures so their adsorption is endothermic (e.g. Figure 2.1). There appeared to be a relation between ΔH and ΔS, but this could well be because the compounds did not differ much in their affinity for the enzyme; values of ΔG all lay within the range 3.3–$4.8\,kcal\,mol^{-1}$ (14–$20\,kJ\,mol^{-1}$).

The results establish the importance of entropy in determining affinity and that a change in structure, such as the addition of a methylene group, produces different effects on the entropy of adsorption in different molecules. It is likely that some of these differences arise from effects on water and Belleau has suggested that drugs might induce conformational changes in receptor

TABLE 5.6: *Estimates of free energies (ΔG), enthalpies (ΔH) and entropies (as $T\Delta S$) of adsorption for compounds to muscarinic acetylcholine receptors of the guinea-pig ileum, made from measurements of dose-ratios at 29° and 37°C.*

	ΔG kcal mol^{-1}	ΔH kcal mol^{-1}	$T\Delta S$ kcal mol^{-1}
$(-)$-Hyoscine	-13.1	-17.1	-4.0
$(-)$-Hyoscine methiodide	-14.0	-36.4	-22.4
$Ph_2CHCH_2OCH_2CH_2\overset{+}{N}Me_3Br^-$	-8.7	-6.4	$+2.3$
$Ph_2CHCOOCH_2CH_2\overset{+}{N}Me_3Br^-$	-9.7	-2.1	$+7.6$
Ph_2CHCOO—⟨$+NMe_2$⟩I^-	-12.5	-17.1	-4.6
		$-9.1*$	$+3.4*$
$Ph_2C(OH)COO$—⟨$+NMe_2$⟩	-13.5	$+8.0$	$+21.5$
3-Diphenylacetoxyquinuclidine methiodide	-10.9	-9.1	$+1.8$
Diphenylacetyl-*pseudo*-tropine methiodide	-11.4	-8.0	$+3.4$

Note that the errors attached to estimates of ΔH and $T\Delta S$ may be as big as $\pm 5\,\text{kcal mol}^{-1}$; the asterisk indicates a second set of results obtained with this compound. The negative values of $T\Delta S$ found with hyoscine methiodide were also found with $(-)$-hyoscyamine methiodide ($\Delta G = -13.5$, $\Delta H = -29.2$, $T\Delta S = -16$) but not with $(+)$-hyoscyamine methiodide ($\Delta G = -10.8$, $\Delta H = -6.7$, $T\Delta S = +4$).

(From results of Barlow, Berry, Glenton, Nikolaou and Soh, *British Journal Pharmacol.* (1976), no. 58, p. 619.)

proteins by their effects on water and these might explain drug action.[20]

It is arguable how far results obtained with acetylcholinesterase can reasonably be extended to the muscarinic receptor and it is difficult to perform similar experiments with tissues such as the isolated guinea-pig ileum, because this can be used only over a narrow temperature range, from about 29–37°C. The errors in measuring dose-ratios are also greater than those in enzymic reactions. It has, nevertheless, been found that there are apparent differences in the effects of temperature on the affinity constants of antagonists.[21] Some results are shown in Table 5.6 and though the estimates of ΔH and $T\Delta S$ are only very approximate, there are differences in the sign of ΔH which are beyond experimental error. Moreover, some changes appear to be associated with chemical structure and are not merely random. More accurate measurements could be made with isolated receptor preparations made from membrane fragments.

Although the interpretation of these entropy effects is obscure, they emphasise the need to consider water in the binding process. The drug in solution is hydrated and so is the receptor. The formation of the complex must involve alterations in water structure and probably the displacement of water molecules in the receptor environment by drug molecules. This has been emphasised by Gill in a review of drug-receptor interactions;[22] these must be thought of as drug-water-receptor interactions.

The importance of water and hydrophobic bonding in determining protein structure is well known.[23] Hydrogen bonds, on the other hand, will be greatly weakened by the presence of water. Though it is possible that the drug may interact with a group in a hydrophobic part of the receptor, the hydrogen bonding group in the drug will interact with water in the bulk phase and these bonds must be broken in any interaction with the receptor.

An idea of the strength of hydrogen bonds can be obtained from the distance between the two electronegative atoms which form donor and acceptor. In the strongest hydrogen bond, $[F-H..F]^-$, the distance between the fluorine atoms is 2.26 Å. In Ice I the distance between two oxygen atoms is 2.76 Å and the bond energy is about 5 kcal mol^{-1} (21 kJ mol^{-1}).[24] Bonds involving the less electronegative nitrogen atom are weaker and in base-pairs formed by adenine and thymine in nucleic acids the N—H..O bond length is 2.82 Å and the N—H..N bond length is 2.91 Å; in base-pairs formed from guanine and cytosine the distances are O—H..N, 2.84 Å; N—H..N, 2.92 Å; N—H..O, 2.84 Å.[25] The individual bonds are weaker than those between water molecules but the combined effect of many bonds is great. The highest observed contribution from a hydroxyl group to affinity at muscarinic acetylcholine receptors (2.7 kcal mol^{-1}, 11.3 kJ mol^{-1}; page 211) seems, therefore, to be reasonable for a hydrogen bond between drug and receptor, though more information is needed about the interaction between the hydroxyl group and water in solution, as well as about the environment of the binding group in the receptor.

Measurements of the affinities of compounds for muscarinic acetylcholine receptors can also be used to obtain some idea of the actual countribution to binding made by coulombic interaction between positive and negative charges. Some compounds have been studied[26] in which quaternary nitrogen, $-\overset{+}{N}Me_3$, is replaced by quaternary carbon, $-CMe_3$, and the results are shown in Table 5.7. The biggest effect observed, with (±)-phenylcyclohexylglycollic acid, indicates a difference in the free energy of adsorption of

TABLE 5.7: *Comparison of affinities of esters of choline and 3,3-dimethylbutanol for muscarinic acetylcholine receptors of the guinea-pig ileum at 37°C. Mean estimates of $\log K$ are shown with the difference ($\Delta \log K$) and the effect on the free energy of adsorption ($\Delta\Delta G$).*

				$\Delta\Delta G$	
R=	$ROCH_2CH_2CMe_3$	$ROCH_2CH_2\overset{+}{N}Me_3$	$\Delta \log K$	kcal mol^{-1}	kJ mol^{-1}
$Ph_2C(OH)CO-$	7.53	8.51	0.98	1.4	5.8
(±) [Ph, OH, C, C$_6$H$_{11}$, COO—]	7.58	9.36	1.78	2.5	10.6
Ph_2CHCO-	5.49	7.16	1.6	2.3	9.5
$(\pm)PhCH(OH)CO-$	5.49	5.29	−0.20	−0.3	−1.3

(From results of Barlow and Tubby, *British Journal Pharmacol.* (1974), no. 51, p. 97.)

2.5 kcal mol^{-1} (10.6 kJ mol^{-1}). Although this is only a small group of comparisons, it confirms the idea that coulombic interactions in water will be weakened by the hydration shells around the charges, though a comparison of $-\overset{+}{\text{N}}\text{Me}_3$ with $-\text{CMe}_3$ may not be a good way of assessing the contribution which the charge makes to binding. Although the $\overset{+}{\text{N}}$—C and C—C bonds are similar in length, the groups do not have the same size in water (see Example 3.1) and the $-\text{CMe}_3$ group may be associated with considerable hydrophobic bonding which could explain why the mandelic ester of 3, 3-dimethylbutanol has higher affinity than the mandelic ester of choline.

Conclusion

Molecular pharmacology is in its infancy. In the past, attempts to understand how drugs may affect receptors have involved arguing backwards from the observed relations between structure and activity. If the estimates of biological activity depend only on binding to receptors, these may give some idea of the chemical features important for binding. Computer methods have even been developed for scanning molecular structures to try to recognise structural similarities. Even with antagonists, however, it is doubtful how effectively this information can be used to deduce the shape of the receptor and the nature of the groups in it.

A direct approach is now becoming possible. With some enzymes and substrates the molecular structure of the complex can be worked out by X-ray diffraction. Binding may also be studied by n.m.r. techniques. The enzyme, however, must be available in a pure state and its structure must be known. Advances depend upon the biochemist and the protein chemist, as well as on the physical chemist. With studies on receptors there is a further problem; most receptors being membrane-bound structures, is it possible to take them off membranes without altering them? With some types of receptor at least it seems likely that this can be done and that it is only a matter of time before studies with purified receptor material provide direct information about drug-receptor interactions. With these systems, for instance, it may be possible to obtain rate constants for binding processes which are not obtainable with intact systems because of the limits imposed by diffusion within tissues.

The next step in the sequence of events after the formation of the agonist-receptor complex in some instances is now known to be the activation of adenylate cyclase inside the cell and the formation of cyclic adenosine monophosphate (or cyclic guanosine monophosphate in some systems). The 'message' is thus transferred from the receptor across the cell wall to the 'second messenger' which acts intracellularly but the details of this link will be extremely difficult to work out quantitatively. To a considerable extent they should account for the shape of the relationship between stimulus and response (Figure 4.8). The necessary information may be obtainable from experiments with cells in tissue culture.

In one sense the rules in molecular pharmacology have still to be worked out but in another sense they are already known. The underlying chemical ideas are simple. It is the complexity of the system which is the problem and makes for uncertainty.

Molecular pharmacology supplements rather than replaces the empirical approach to the development of new drugs for the treatment of disease. It is not yet possible to design drugs in a purely logical fashion. In 1940, when Woods[27] discovered that *p*-aminobenzoic acid is an essential metabolite for certain bacteria and that sulphonamides act as antimetabolites, Fildes[28] appealed for a rational approach to chemotherapy, based on an understanding of the biochemistry of parasite and host. In fact, however, the subsequent spectacular developments have almost all come from screening tests looking for naturally occurring antibacterial substances (antibiotics) produced by other micro-organisms. It is these substances which have led to advances in understanding bacterial biochemistry, rather than the other way round. Attempts to correlate chemical structure quantitatively with biological activity, such as have been described in this chapter, have been very valuable in drawing attention to the need to consider physicochemical properties in drug design and not simply the ease of synthesis. It would be incorrect, however, to claim that drugs are now designed rationally. To a large extent the process is still empirical.

It is important for the chemist to recognise that new discoveries in drug research originate from biological observations—chemistry is no substitute for biology—but it is also important that the biologist should appreciate that fundamentally his problems are chemical. There is the dilemma, though, that any scientist must apply Occam's razor and work with the simplest hypothesis even though this is likely to be inadequate when there is more information. It is therefore necessary to be cautious when applying simple ideas to a complex situation. As A. J. Clark wrote:[29] 'It seems fair to assume as a general principle that if a pharmacological reaction appears simpler than an analogous reaction in non-living systems, the simplicity must be apparent rather than real...' But it was Clark, more than anyone at that time (1933) who pointed out the need for applying chemical ideas to drug action.[30] The big advances which have since been made owe much to his influence but the situation is the same today. It calls for perseverance in thinking in molecular terms together with the recognition that the complex nature of most biological problems demands caution in interpreting them in this way.

References

1. K. H. Meyer and H. Hemmi, *Biochim. Z.* (1935), no. 277, p. 39.

2. J. Ferguson, *Proc. Roy. Soc., B* (1939), no. 127, p. 387.

3. A. Frumkin, *Z. Physik. Chim.* (1925), no. 116, p. 501; R. Collander, *Acta Physiol. Scand.* (1947), no. 13, p. 363.

4. R. Collander, *Acta Chem. Scand.* (1954), no. 5, p. 774.

5. C. Hansch, P. P. Maloney, T. Fujita and R. M. Muir, *Nature* (1962), no. 194, p. 178.

6. C. Hansch, R. M. Muir, T. Fujita, P. P. Maloney, F. Geiger and M. Streich, *Journal American Chem. Soc.* (1963), no. 85, p. 2817.

7. C. Hansch and T. Fujita, *Journal American Chem. Soc.* (1964), no. 86, p. 1616.

8. J. Iwasa, T. Fujita and C. Hansch, *Journal Med. Chem.* (1965), no. 8, p. 150.

9. C. Hansch and S. M. Anderson, *Journal Med. Chem.* (1967), no. 10, p. 745.

10. C. Hansch, A. R. Steward, J. Iwasa and E. W. Deutsch, *Mol. Pharmacol.* (1965), no. 1, p. 205; C. Hansch, A. R. Steward, S. M. Anderson and D. Bentley, *Journal Med. Chem.* (1968), no. 11, p. 1.

11. M. S. Tute, in *Advances in Drug Research*, N. J. Harper and A. B. Simmonds (eds.) (Academic Press, New York, 1971), p. 1.

12. C. Hansch, in *Chemical Structure and Biological Activity*, J. A. Keverling Buisman (ed.) (Elsevier, Amsterdam, 1977), p. 47, p. 292.

13. R. B. Barlow, K. A. Scott and R. P. Stephenson, *British Journal Pharmacol.* (1963), no. 21, p. 509.

14. S. M. Free and J. W. Wilson, *Journal Med. Chem.* (1964), no. 7, p. 395.

15. F. B. Abramson, R. B. Barlow, M. G. Mustafa and R. P. Stephenson, *British Journal Pharmacol.* (1969), no. 37, p. 207.

16. F. B. Abramson, R. B. Barlow, F. M. Franks and J. D. M. Pearson, *British Journal Pharmacol.* (1974), no. 51, p. 81.

17. R. B. Barlow, F. M. Franks and J. D. M. Pearson, *Journal Med. Chem.* (1973), no. 16, p. 439.

18. B. Belleau, H. Tani and F. Lie, *Journal American Chem. Soc.* (1965), no. 87, p. 2283.

19. B. Belleau and J. L. Lavoie, *Canadian Journal Biochem.* (1968), no. 46, p. 1397.

20. B. Belleau, *Journal Med. Chem.* (1964), no. 7, p. 776; *Ann. NY Acad. Sci.* (1967), no. 144, p. 705; in *Physicochemical Aspects of Drug Action*, E. J. Ariëns (ed.) (Pergamon, Oxford, 1968), p. 207.

21. R. B. Barlow, K. J. Berry, P. A. M. Glenton, N. M. Nikolaou and K. S. Soh, *British Journal Pharmacol.* (1976), no. 58, p. 613.

22. E. W. Gill, in *Progress in Medicinal Chemistry*, vol. 4, G. P. Ellis and G. B. West (eds.) (Butterworth, London, 1965), p. 39.

23. C. Tanford, *The Hydrophobic Effect* (Wiley, New York, 1973); F. Franks, in *Water, A Comprehensive Treatise*, vol. 4 (Plenum Press, New York, 1975), p. 1; A. Suggett, in *Biological Activity and Chemical Structure*, J. A. Keverling Buisman (ed.) (Elsevier, Amsterdam, 1977), p. 95.

24. L. Pauling, *The Chemical Bond* (Cornell University Press, Ithaca, 1967), p. 228; J. C. Speakman, *The Hydrogen Bond* (Chemical Society, London, 1975).

25. S. Arnott, M. H. F. Wilkins, L. D. Hamilton and R. Langridge, *Journal Mol. Biol.* (1965), no. 11, p. 391.

26. R. B. Barlow and J. H. Tubby, *British Journal Pharmacol.* (1974), no. 51, p. 95.

27. D. D. Woods, *British Journal Exp. Path.* (1940), no. 21, p. 74.

28. P. Fildes, *Lancet* (1940), no. i, p. 955.

29. A. J. Clark, *The Mode of Action of Drugs on Cells* (Arnold, London, 1933).

30. A. J. Clark, *Handbuch der Experimentellen Pharmakologie*, IV (Springer, Berlin, 1937) (in English, reprinted 1973).

Appendix: Line-fitting by the method of least-squares

A Straight Line

The equation representing a straight line is:

$$y = mx + c$$

A least-squares fit of y on x calculates the values of m and c such that the sum of the values of $(y_{observed} - y_{calculated})^2$ i.e. $S(y - mx - c)^2$, is a minimum, where y and x are the experimental values obtained (Figure A.1). When this is multiplied out the sum of squares is:

$$Sy^2 - 2mSxy - 2cSy + 2mcSx + m^2Sx^2 + nc^2$$

where Sy^2 is the sum of the n values of y^2 etc.

The values of m and c for which this total is a minimum can be calculated by calculus; for a minimum the partial differential coefficients of the sum with respect to m and c should be zero, i.e.

$$-2Sxy + 2cSx + 2mSx^2 = 0 \text{ (differentiating with respect to } m) \text{ and}$$

$$-2Sy + 2mSx + 2nc = 0 \text{ (differentiating with respect to } c)$$

So $mSx^2 + cSx = Sxy$ and $mSx + nc = Sy$. So

$$m = \frac{nSxy - SxSy}{nSx^2 - (Sx)^2}$$

and

$$c = \frac{SxySx - SySx^2}{(Sx)^2 - nSx^2} \quad \text{or} \quad \frac{Sy - mSx}{n}$$

For each set of values of x and y the values of x^2 and xy are also calculated and the totals Sx, Sy, Sx^2 and Sxy are obtained and m and c calculated from these. The summing operations can very easily be performed by a computer or desk calculator.

Note: Sx^2 is the sum of the values of x^2; it is not the same as $(Sx)^2$ which is the square of the sum of the values of x.

224

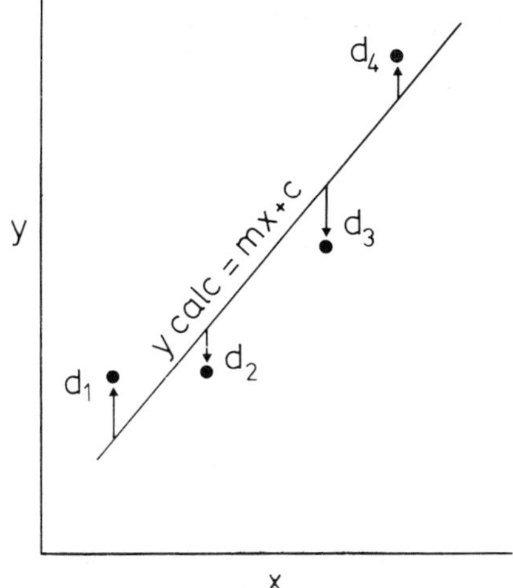

FIGURE A.1 Fitting results to a straight line by the method of least-squares. The values of x and y are related by the expression $y_{calc} = mx + c$ with the values of m and c such that the sum of $(y - y_{calc})^2$, i.e. $d_1^2 + d_2^2 + d_3^2 + d_4^2$, is a minimum. This is a fit of y to x; if x is fitted to y different values of m and c will be obtained.

Having calculated m and c, it is desirable to know how good the fit is. This can be judged by comparing the sum of the squares of the difference, between the observed and calculated values, $S(y_{obs} - y_{calc})^2$, with the total sum of squares, $S(y_{obs})^2$. The sum of squares of the difference can be obtained in one of three ways. It can be calculated from the values of Sy^2, Sxy, Sy, Sx and Sx^2 from the expression shown above, once m and c are known. If the mean values, $\bar{x}$ and $\bar{y}$, are calculated,

$$S(y_{obs} - y_{calc})^2 = S(y - \bar{y})^2 - \frac{[S(x - \bar{x})(y - \bar{y})]^2}{S(x - \bar{x})^2}$$

and it can also be obtained by using the values of m and c to calculate y for each value of x, and the squares of the difference from the observed value can then be added up directly, a simple operation for a computer.

The variance of the values of x about the mean ($\bar{x}$) is

$$s_x^2 = \frac{S(x - \bar{x})^2}{n - 1} = \frac{nSx^2 - (Sx)^2}{n(n - 1)}$$

and similarly the variance of the values of y about the mean,

$$s_y^2 = \frac{S(y - \bar{y})^2}{n - 1} = \frac{nSy^2 - (Sy)^2}{n(n - 1)}.$$

The association of x with y can be measured by the covariance, C_{xy}, which is

$$\frac{S(y-\bar{y})(x-\bar{x})}{n-1} = \frac{S(xy-y\bar{x}-x\bar{y}+\bar{x}\bar{y})}{n-1}$$

$$= \left(Sxy - Sy\frac{Sx}{n} - Sx\frac{Sy}{n} + n\frac{Sx}{n}\frac{Sy}{n}\right)\Big/(n-1)$$

$$= \frac{nSxy - SySx}{n(n-1)}.$$

The product moment correlation coefficient,

$$r = \frac{C_{xy}}{(s_x^2 s_y^2)^{1/2}} = \frac{nSxy - Sx\,Sy}{\{[nSx^2 - (Sx)^2][nSy^2 - (Sy)^2]\}^{1/2}}$$

and this should be 1 if the points lie exactly on a line and y increases as x increases, or -1 if the points lie exactly on a line and y decreases as x increases.

For example, the binding of benzylpenicillin to bovine serum albumin (Table 4.2, page 125) gives the following values of x and y for the Scatchard plot (Figure 4.2)

x	y
31	5.81
43	5.35
116	3.97
143	3.92
232	3.66
309	3.37
522	2.68
564	2.69

and $Sx = 1960$; $Sx^2 = 776600$; $Sy = 31.45$; $Sy^2 = 132.68$; $Sxy = 6237.8$, so

$$m = \frac{8(6237.8) - 1960(31.45)}{8(776600) - 1960(1960)} = -0.00495$$

and

$$c = \frac{31.45 - (-0.00495)1960}{8} = 5.144.$$

The correlation coefficient,

$$r = \frac{8(6237.8) - 1960(31.45)}{\sqrt{[8(776600) - (1960)^2][8(132.68) - (31.45)^2]}} = -0.8965$$

The sum of the squares of the difference $S(y_{obs} - y_{calc})^2$ is

$$132.68 - 2(-0.00495)6237.8 - 2(5.144)31.45$$

$$+ 2(-0.00495)5.144(1960) + (-0.00495)^2 (776600)$$

$$+ 8(5.144)^2 = 1.77,$$

compared with $Sy^2 = 132.68$. If the first two results, which lie well off the line, are omitted, $m = -0.00304$, $c = 4.337$ and $r = -0.997$. The sum of the squares of the difference $= 0.01$, compared with $Sy^2 = 70.30$. With these 6 points there is a close fit to a straight line and a high negative correlation. With all 8 points the fit is not so good.

For the calculation of the variance of the scatter of the points about the line there are $n-2$ degrees of freedom so

$$s^2 = \frac{S(y_{obs} - y_{calc})^2}{n-2}$$

where s^2 represents the residual variance, for scatter unaccounted for by the regression of y with x. The variance of the slope, m, is given by

$$\frac{s^2}{S(x - \bar{x})^2} = \frac{ns^2}{(nSx^2 - (Sx)^2)}$$

and the variance of the constant, c, is given by $s^2 Sx^2/[nSx^2 - (Sx)^2]$. The t-distribution can be used to set limits to the estimates of m and c.[1,2]

The Hansch Type Polynomial

Results can be fitted by least-squares to polynomial expressions in a similar way to obtain directly the coefficients but there is an additional equation to be solved for each extra term involved. The Hansch type of polynomial is an example:

$$y_{calc} = a\pi^2 + b\pi + c\sigma + d$$

$$(y - y_{calc})^2 = a^2\pi^4 + 2a(b\pi^3 + c\pi^2\sigma + d\pi^2 - y\pi^2) + b^2\pi^2$$

$$+ 2b(c\pi\sigma + d\pi - y\pi) + c^2\sigma^2 + 2c(d\sigma - cy\sigma)$$

$$+ d^2 - 2dy + y^2$$

If this is summed for n results,

$$S(y - y_{calc})^2 = a^2 S\pi^4 + 2a(b S\pi^3 + c S\pi^2\sigma + d S\pi^2 - Sy\pi^2) + b^2 S\pi^2$$

$$+ 2b(c S\pi\sigma + d S\pi - Sy\pi) + c^2 S\sigma^2 + 2c(d S\sigma - c Sy\sigma)$$

$$+ nd^2 - 2d\, Sy + Sy^2$$

For a minimum the partial differential coefficients with respect to a, b, c and d equal zero so:

$$aS\pi^4 + bS\pi^3 + cS\pi^2\sigma + dS\pi^2 = Sy\pi^2$$

$$aS\pi^3 + bS\pi^2 + cS\pi\sigma + dS\pi = Sy\pi$$

$$aS\pi^2\sigma + bS\pi\sigma + cS\sigma^2 + dS\sigma = Sy\sigma$$

$$aS\pi^2 + bS\pi + cS\sigma + dn = Sy$$

and values of a, b, c, and d can be calculated from these four equations.

For each experimental result (y) there will be a corresponding value of π and σ so the computer calculates π^2, π^3, π^4, $\pi\sigma$, $\pi^2\sigma$, σ^2, $y\pi^2$, $y\pi$ and $y\sigma$, and keeps a running total of these values, together with totals for π, σ and y. Some of these numbers may be very small and in the multiplication necessary to solve the four equations for the four unknowns, a, b, c and d, it may be necessary to check that they are not too small for the computer to handle. If there is any risk of this, both sides of the equations can be multiplied by a large number or a programme written in Fortran can specify 'double precision'.

The goodness of fit can be assessed from the residual variance which is given by $S(y_{obs} - y_{calc})^2$ and from the correlation coefficient, **r**, obtained by using the observed and calculated values as **x** and **y** in a simple linear regression.

Hyperbola. The equation representing a hyperbola is:

$$y = \frac{Mx}{x + K}.$$

In this form y may represent the rate of an enzymic reaction, or the amount of adsorbed material, and x the concentration of substrate or substance adsorbed; M corresponds to the maximum rate or amount adsorbed and K to the Michaelis constant or dissociation constant of the adsorbed complex. It is not possible to obtain the conditions for a minimum in the sum of $(y - y_{calc})^2$ directly, as was done with the Hansch type of equation, but an approximate answer can be obtained by applying Taylor's theorem and the values of M and K obtained can then be used to obtain more accurate answers by repeating the calculations. The reasoning behind the approximation is discussed by Snedecor and Cochran,[3] and in more detail by Parker and Waud.[4] The partial differential coefficients, $x/(x+K)$ and $-x/(x+K)^2$, must be calculated for each value of x, so an initial estimate of K must be made. Each value of x is converted into these coefficients and if these are x_1 and x_2, respectively, it is necessary to calculate x_1^2, $x_1 x_2$, x_2^2, $x_1 y$ and $x_2 y$. The sums of these values are related to parameters B_1 and B_2 by the equations:

$$Sx_1^2(B_1) + Sx_1 x_2(B_2) = Sx_1 y$$
$$Sx_1 x_2(B_1) + Sx_2^2(B_2) = Sx_2 y$$

These can be solved for B_1 and B_2 and the value of K is increased by B_2/B_1. The values of x_1 and x_2 are recalculated and the whole process repeated until changes in K are considered to be negligible. The value of B_1 corresponds to M.

The initial estimate of K must be such that the calculations can converge but the concentration which produces half-saturation is sure to be suitable. Difficulties only occur when the results do not indicate some degree of saturation, i.e. when only the top or the bottom parts of the saturation curve have been studied, and these results will not be suitable for calculating the Michaelis constant by any method.

The calculations can be made with quite small computers. The following programme has been used with a PDP 8L machine with a 4K core.

Program:

```
C-FOCAL,1969

01.10 T "  HYPMIC"

02.05 A !,"NUMBER OF PAIRS OF RESULTS"N
02.10 FOR I=1,1,N; DO 12.2
02.30 A !,"EST K"K
02.40 FOR J=1,1,6; DO 3.0
02.42 S SY=0; S SD=0
02.43 T !,"    CONC        OBS RATE    CALC RATE    DIFF        Y/(1-Y)"
02.45 FOR I=1,1,N; DO 5.0
02.47 T !, "DIFF SQ"SD, "TOTAL SQ" SY
02.48 T !,%8.04 "MICHAELIS CONSTANT"K
02.50 QUIT

03.05 S S1=0; S S2=0; S S3=0; S S4=0; S S5=0
03.10 FOR I=1,1,N; DO 4.0
03.15 S B2=(S4*S2-S5*S1)/(S2*S2-S1*S3)
03.20 S B1=(S4*S3-S5*S2)/(S1*S3-S2*S2)
03.25 S K=K+B2/B1
03.30 T !,%8.04," K"K," B1(MAX RATE)"B1," B2"B2

04.05 S X1=X(I)/(X(I)+K)
04.06 S X2=-X1/(X(I)+K)
04.10 S S1=S1+X1*X1; S S2=S2+X1*X2; S S3=S3+X2*X2
04.15 S S4=S4+X1*Y(I); S S5=S5+X2*Y(I)

05.10 S R=B1*X(I)/(X(I)+K)
05.15 S D=Y(I)-R; S SD=SD+D*D; S SY=SY+Y(I)*Y(I)
05.17 S H=Y(I)/(B1-Y(I))
05.20 T !,%8.04 X(I), Y(I), R, D, H

12.20 A !,"CONC" X(I); A "RATE" Y(I)
*
```

Example: results obtained with acetylthiocholine (mM concentration) and acetylcholinesterase.

```
   G
    HYPMIC
 NUMBER OF PAIRS OF RESULTS:8
 CONC:.05 RATE:15.5
 CONC:.1 RATE:24
 CONC:.1 RATE:24.25
 CONC:.15 RATE:29
 CONC:.2 RATE:32
 CONC:.2 RATE:31.75
 CONC:.3 RATE:35.
 CONC:.4 RATE:37
 EST K:.1
    K=       0.0902 B1(MAX RATE)=      45.8499 B2=-      0.4479
    K=       0.0912 B1(MAX RATE)=      45.9524 B2=       0.0452
    K=       0.0912 B1(MAX RATE)=      45.9480 B2=-      0.0011
    K=       0.0912 B1(MAX RATE)=      45.9483 B2=       0.0001
    K=       0.0912 B1(MAX RATE)=      45.9481 B2=-      0.0001
    K=       0.0912 B1(MAX RATE)=      45.9481 B2=       0.0001
         CONC          OBS RATE     CALC RATE      DIFF        Y/(1-Y)
  =      0.0500=      15.5000=      16.2715=-     0.7715=      0.5091
  =      0.1000=      24.0000=      24.0324=-     0.0324=      1.0935
  =      0.1000=      24.2500=      24.0324=      0.2176=      1.1176
  =      0.1500=      29.0000=      28.5756=      0.4244=      1.7111
  =      0.2000=      32.0000=      31.5586=      0.4414=      2.2942
  =      0.2000=      31.7500=      31.5586=      0.1914=      2.2362
  =      0.3000=      35.0000=      35.2370=-     0.2370=      3.1969
  =      0.4000=      37.0000=      37.4176=-     0.4176=      4.1349
 DIFF SQ=      1.2857TOTAL SQ= 6871.3800
 MICHAELIS CONSTANT=      0.0912*
```

The Logistic Expression. The logistic expression:

$$y = M \frac{x^p}{x^p + K^p}$$

can be used empirically to fit response to log dose and the procedure is very similar to that for the hyperbola except that there is an additional parameter (p). The partial differential coefficients which must be calculated are

$$x_1 = \frac{x^p}{x^p + K^p}; \quad x_2 = \frac{-x^p(pK^{p-1})}{(x^p + K^p)^2}$$

and

$$x_3 = \frac{x^p K^p \ln (x/K)}{(x^p + K^p)^2}$$

and the parameters B_1, B_2 and B_3 are related by the equations:

$$Sx_1^2(B_1) + Sx_1x_2(B_2) + Sx_1x_3(B_3) = Sx_1y$$

$$Sx_1x_2(B_1) + Sx_2^2(B_2) + Sx_2x_3(B_3) = Sx_2y$$

$$Sx_1x_3(B_1) + Sx_2x_3(B_2) + Sx_3^2(B_3) = Sx_3y$$

Starting values of K and P are used for the calculations of x_1, x_2 and x_3 and from the summed values of x_1, x_1^2, $x_1 x_2$, x_2, x_2^2, $x_1 x_3$, $x_2 x_3$, x_3, x_3^2, $x_1 y$, $x_2 y$, and $x_3 y$ the values of B_1, B_2 and B_3 are calculated. The value of K is now increased by B_2/B_1 and that of p by B_3/B_1 and the process is repeated until changes in these are negligible.

For results obtained in experiments where the graphs of response against log dose are thought to be parallel it is possible to fit numbers of lines to logistic expressions with the same slope and maximum and the ratios of the parameter K can be used to calculate potency ratios or dose-ratios.[5]

Other Methods. For expressions where the coefficients in a least-squares fit cannot be obtained directly, as with a polynomial, or by applying Taylor's theorem, it may be possible to use a search technique which calculates how $S(y_{obs} - y_{calc})^2$ alters as the various coefficients are altered, using particular starting values and steps of a particular size, i.e. doing the partial differentiation by trial rather than by calculus. These may provide answers when other methods fail but they take a lot of computing time and require care in operating.

Note: The least squares procedures described in this Appendix can easily be performed with the recently produced micro-computers; I use an 8K Commodore PET 2001.

References

1. R. C. Campbell, *Statistics for Biologists* (Cambridge University Press, 1967). p. 195.

2. D. Colquhoun, *Lectures on Biostatics* (Oxford University Press, 1971), p. 214.

3. G. W. Snedecor and W. G. Cochran, *Statistical Methods* 6th edn (Ames, Iowa, 1967), p. 466.

4. R. B. Parker and D. R. Waud, *Journal Pharmacol.* (1971), no. 177, p. 1.

5. D. R. Waud and R. B. Parker, *Journal Pharmacol.* (1971), no. 177, p. 13; R. B. Barlow, *British Journal Pharmacol.* (1975), no. 53, p. 139.

Answers to Examples

Chapter 1

EXAMPLE 1:1. $\bar{r} = 5$; $d = -2,\ -1,\ 0,\ 1,\ 2$; $Sd^2 = 10$; $Sr^2 = 135$; $n\bar{r}^2 = 125$.

EXAMPLE 1:2. Means: 6.991; 7.094. Difference $= 0.103$: standard errors, 0.004 (7); 0.007 (6); combined $se = 0.008$: t for 11 degrees of freedom, 2.21 $(P = 0.05)$, 3.12 $(P = 0.01)$, so the difference is significant at the lower level of probability.

EXAMPLE 1:3.

1. Rank sums: males   88 $U_1 = 100 + 55 - 122 = 33$
                    females   122 $U_2 = 100 + 55 - 88\ \ = 67$
    Not significant $(P = 0.05)$, Table 1.3.

2. Rank sums: controls   442 $U_1 = 400 + 210 - 378 = 232$
                    placebo   378 $U_2 = 400 + 210 - 442 = 168$
    Not significant $(P = 0.05)$

3. Rank sums: controls   325 $U_1 = 115$
                    alcohol $U_2 = 285$
    Not significant $(P = 0.05)$ but very nearly, the limiting value is 113.

4. Rank sums: males   77 $U_1 = 22$
                    females   133 $U_2 = 78$
    Significant $(P = 0.05)$

EXAMPLE 1:4. Concentration $= 50 \times 1.5 \times 0.3\ \text{mM} = 22.5\ \text{mM}$.

EXAMPLE 1:5. Concentration

$$= \frac{0.15}{0.2} \times 0.926 \times 10^{-5}\,\text{M} = 0.69 \times 10^{-5}\,\text{M}\ (= 6.9\,\mu\text{M}).$$

EXAMPLE 1:6. Means:

$$\text{Low standard} = 19.375, \quad \text{High standard} = 35.624$$
$$\text{Low unknown} = 18.50, \quad \text{High unknown} = 34.375$$

$$V = 6.741; \quad B = 74.4; \quad b = 53.36; \quad M = -0.020; \quad g = 0.124.$$

Correction factor $= 0.955$ so

$$\text{concentration} = 0.955 \times \frac{0.1}{0.15} \times 5 \times 10^{-5} = 3.18 \times 10^{-5}\,\text{M};$$

95% confidence limits 73% and 123% of this, i.e. 23.2 and 39.1 μM. The limits are large because the variance is large; it can be seen that the sensitivity of the preparation is increasing during the assay. There is no significant deviation from parallelism.

EXAMPLE 1:7. The concentration in the dilute extract is estimated to be 3.38 μM so the concentration in 1 g of nettle leaf, assuming it to be 1 ml is 33.8 mM = 6.13 mg acetylcholine chloride/ml.

If the volume of liquid in 500 hairs is 4.5 μl, the concentration of acetylcholine chloride is 11.1 mg/ml. The volume of fluid in the leaf will be less than 1 ml/g, nor is it likely that the acetylcholine is uniformly distributed.

The estimates with individual hairs do not indicate any significant differences ($P = 0.05$) between the test preparation or between hairs from different parts of the plant.

EXAMPLE 1:8. Mean lethal doses: 18.33 and 10.00; potency = 1.88 units/ml. Mean log lethal doses: 1.272 and 0.998; potency = 1.86 units/ml. The combined standard error is 0.024 and with $t = 2.23$ for $P = 0.05$ and 10 degrees of freedom, the log limits are ± 0.054 and the 95% confidence limits are $1.86/1.13 = 1.65$ units/ml and $1.86 \times 1.13 = 2.10$ units/ml. If doses are used rather than log doses it is necessary to deal with the variance of a ratio of two means, which has not been considered in this book.

EXAMPLE 1:9.

$$M = \frac{0.8004}{2.7588} \times 0.2218 = 0.0643;$$

potency $= 1.16 \times 28 = 32.5$ milliunits in the volume given.
The mean weight, $\bar{w} = 0.53$ and the variance,

$$V = \frac{1}{24 \times 0.53} = 0.0786.$$

G is 0.273 and there is no significant deviation from parallelism

$$\left(\frac{G}{2(V)^{1/2}} = 0.49 \right).$$

$F = 0.4002$; $b = 6.219$; $B = 1.598$; $g = 0.159$ and the log confidence limits are

$$0.0643 + 0.0122 \pm 0.3747(0.0727)^{1/2} = 0.0765 \pm 0.101,$$

i.e. -0.0245 and 0.1775; the 95% confidence limits are $0.945 \times 28 = 26.5$ and $1.505 \times 28 = 42.1$ milliunits in the volume given.

EXAMPLE 1:10. Frog rectus estimate: $1.414 \times 2.5 \times 10^{-5} = 35.35\,\mu M$. Guinea-pig ileum estimate: $20 \times 5\,\mu M = 100\,\mu M$. This difference is suspicious (the index of discrimination is 2.8).

EXAMPLE 1:11. The approximate dose-ratio is 4; $M = 0.040$ so the exact ratio is estimated to be $4 \times 1.098 = 4.39$.

EXAMPLE 1:12. The approximate dose-ratio is 20; $M = -0.073$ so the exact ratio is $20 \times 1.18 = 23.7$.

Chapter 2

EXAMPLE 2:1. $32.5\,\mu g$ (this should not produce any ill-effects but mouth-pipetting of solutions of cyanide should be avoided).

EXAMPLE 2:2. $73.5\,mg$ of atropine sulphate contain $0.2116\,mmol$ of atropine which should be dissolved in $21.16\,ml$ to obtain a $10\,mM$ solution.

EXAMPLE 2:3. The concentration is $498.4/162.2 = 3.07$ times what it should be.

EXAMPLE 2:4.

$$\frac{22.4}{270} \times 30 \times \frac{310}{273} = 2.83 \text{ litres};$$

the toxic dose in man is around 50 mg and upwards.

EXAMPLE 2:5.

$$\ln K = -\frac{2000}{613.8} = -3.26; \quad K = 0.0384.$$

If the concentrations of ATP and ADP are the same, the ratio $CP/C = 0.0384 \ (=1/26)$.

$$\text{ADP} = \frac{100 \times 3 \times 0.0384}{20} = 0.576 \text{ mM}.$$

If ΔG is zero, $K = 1$ and $\text{ADP} = 15 \text{ mM}$.

EXAMPLE 2:6.   At 37°C $\Delta G = -13.057 \text{ kcal mol}^{-1}$ $(54.58 \text{ kJ mol}^{-1})$

at 29°C $\Delta G = -13.164 \text{ kcal mol}^{-1}$ $(55.03 \text{ kJ mol}^{-1})$

$\Delta \ln K = -2.383 \times 0.320 = -0.737$:

$$\Delta\frac{1}{T} = \frac{1}{310} - \frac{1}{302} = -8.55 \times 10^{-5}$$

$\Delta H = -8.62 \times 10^3 \times R = -17.1 \text{ kcal mol}^{-1}$ $(23.9 \text{ kJ mol}^{-1})$

$T\Delta S = \Delta H - \Delta G = -17.1 + 13.1 = -4.0 \text{ kcal mol}^{-1}$
$$(-16.7 \text{ kJ mol}^{-1}).$$

The negative sign indicates that the binding process is associated with an increase in the order of the system.

EXAMPLE 2:7. $k = 0.092 \text{ h}^{-1}$; $t_{1/2} = 7.53 \text{ h}$; $C_0 = 32.3 \text{ mg l}^{-1}$ so volume of dilution $= 30.96 \text{ l}$. With these results it is not necessary to plot a graph because the increment in $\ln C$ is regular.

EXAMPLE 2:8.   A least-squares fit of $\ln C$ to time gives $k = 0.027 \text{ min}^{-1}$; $t_{1/2} = 25.7 \text{ min}$, $C_0 = 10.2 \text{ units/ml}$, volume of dilution $= 4.91$ for penicillin alone and $k = 0.018 \text{ min}^{-1}$,

$t_{1/2} = 38.5$ min, $C_0 = 12.4$ units/ml, volume of dilution $= 4.0$ l. The excretion of penicillin is reduced by the probenecid, which competes for the transport processes in the kidney which are involved. The volume of dilution represents the extracellular fluid in the dog ($4/20 = 20\%$).

EXAMPLE 2:9. $Q = 0.0625$; $k_r = 0.0866\,\text{h}^{-1}$, $P = 0.25$, when $k_r t = 0.018$ and 1.45, i.e. from 0.2 hrs to 16.75 hrs. Peak absorption occurs when $k_r t = 0.185$, i.e. after 2.14 hrs and $P = 0.831$. After 8 h, $k_r t = 0.69$ and the maximum for the second dose will occur when $k_r t = 0.69 + 0.185 = 0.875$ and P from the initial dose will be 0.44, which together with 0.831 from the second dose will exceed the tolerable level and the subject will not survive.

EXAMPLE 2:10. C_0 for plasma calculated from the results at the start is 85.3 mg/kg and for brain it is 90.3 mg/ml, indicating dilution in 47% and 44% respectively of the body mass. Calculated from the results at the end the values are 17.3 mg/kg for plasma and 45.6 mg/kg for brain, corresponding to dilution in 231% and 88% of the body mass, respectively. Little metabolism should have occurred in this time and the absurdly low value for plasma indicates the uptake by fat. The substance has a high lipid partition coefficient and the drug will effectively be dissolved in total body fluid + the partition coefficient times the mass of fat. The redistribution can also be seen from the fall in brain levels.

EXAMPLE 2:11.   Activity $= 1.42 \times$ proportion   ionised;   at   pH $6.6 = 1.29$,   pH $7.0 = 1.13$,   pH $7.6 = 0.71$,   at   pH $8.0 = 0.44$,   at pH $8.6 = 0.13$.

EXAMPLE 2:12. If C_s is concentration in stomach and C_b is concentration   in   blood, $0.2\,C_s + 50\,C_b = 50$.   At   pH 1.6, $C_s = U(10^7 + 1)$ and C_b at   pH $7.6 = U(10 + 1)$,   where U is concentration unionised so $C_s = 10^6 C_b$ and $C_b = 2.5 \times 10^{-4}$ mg/ml. At pH 6.6,

$$\frac{C_s}{C_b} = \frac{101}{11} \quad \text{and} \quad C_b = 0.964 \,\text{mg/ml};$$

at pH 7.6,

$$C_b = \frac{50}{50.2} = 0.996 \,\text{mg/ml}.$$

EXAMPLE 2:13. (i) at pH 6.5, total concentration in bath $= 1.1 \times 10^{-2}$M, at pH 7.5 it is 2×10^{-3}M and at pH 8.5 it is 1.1×10^{-3}M.

(ii) at pH 6.5 it is 1.1×10^{-3}M, at pH 7.5 it is 2×10^{-3}M and at pH 8.5 it is 1.1×10^{-2}M. Note the effect of the nerve sheath!

EXAMPLE 2:14. σ values: *m*-Cl, 0.34; *m*-nitro, 0.72; *m*-Me, -0.10; *p*-NH$_2$, -0.75.

$$3.52 = 0.34\,a + b$$
$$4.70 = -0.1\,a + b$$
$$a = -2.68, \ b = 4.43 \ (a = \rho, \ b = \log K_0).$$

Value for *m*-nitraniline $= 2.50$ (2.46), for *p*-phenylene diamine $= 6.44$ (6.20).

For the phenols $\rho = 2.82$ and pK_a *m*-cresol $= 9.70$, *m*-nitrophenol $= 12.01$, *p*-aminophenol $= 7.86$. In the anilines and phenols, where the ionising groups are attached more closely to the aromatic ring than in benzoic acid (which has a carbonyl group separating them), the effects of groups are much bigger, as can be seen from the size of ρ.

Values for many *o*-compounds do not fit the general trend because of interactions between the substituent and the ionising groups. With the *m*- and *p*-compounds the least-squares fit gives the equation pK_a (aniline) $= 2.73\,pK_a$ (acid) $- 7.08$ and the correlation coefficient is 0.93: for the *m*-compounds alone the equation is pK_a (aniline) $= 1.99\,pK_a$ (acid) $- 4.06$ and the correlation coefficient is 0.94.

EXAMPLE 2:15. $\Delta H = -4.3$ kcal mol^{-1} (17.8 kJ mol^{-1}); $T\Delta S = 9.0$ kcal mol^{-1} (37.5 kJ mol^{-1}). There is a large increase in entropy associated with ionisation.

EXAMPLE 2:16. Using π values for phenols $\log P$ for *m*-chlorophenol $= 2.50$, *m*-iodophenol $= 2.93$ and resorcinol $= 0.80$ and the values of P are 319, 855 and 6.3, respectively. For the phenoxyacetic acids the value of $\rho = 1.21$ and $\log P_0 = 1.33$, so the partition coefficient of phenoxyacetic acid $= 21$. Using π values for phenoxyacetic acids the values of $\log P$ for the *m*-iodo and *m*-hydroxy compounds are 2.72 and 0.74 and the values of P are 529 and 5.4, respectively.

EXAMPLE 2:17.

$$\Delta R_M = -0.85\pi + 0.59.$$

The point furthest off the line is for the 4-CH_3 compound, for which ΔR_M is 0.34 less than the value given by the equation.

Chapter 3

EXAMPLE 3:1.   Methanol   40.7.   Methylacetate   79.8. ΔV_m for acetate $= 39.1\,cm^3\,mol^{-1}$.   Ethanol   58.7.   Ethylacetate   98.5. $\Delta V_m = 39.8$. Dimethylbutanol 126.3. Dimethylbutylacetate 166.6. $\Delta V_m = 40.3$.

The increment for acetate from ϕ_v^0 for choline bromide and acetylcholine bromide is $33.9\,cm^3\,mol^{-1}$, so the acetyl group is able to fit into the structure of water without apparently occupying as much space as in the liquid state. The values for choline bromide and for acetylcholine bromide indicate that in solution in water they only occupy slightly more space than dimethylbutanol and dimethylbutylacetate, respectively, in the liquid state, even though they contain the additional relatively large bromide ion.

EXAMPLE 3:2. The pairs of values for each wavelength are not significantly different $(P = 0.05)$ with student's t-test but the chances of obtaining consistently smaller values with one solution at all 6 wavelengths are 1 in $2^6 = 1$ in 64 (i.e. less than 0.05). The molar rotations and estimates of stereochemical purity are:

nm			
589	$22.0°$	$-24.4°$	0.951
400	$23.6°$	$-25.4°$	0.965
360	$6.8°$	$-7.4°$	0.959
340	$-15.2°$	$15.8°$	0.981
300	$-148°$	$150°$	0.993
260	$-914°$	$943°$	0.985

The specific rotation $\alpha = 100/\text{m.wt}\,[M] = 14.7°$ and $-16.3°$.

Note the inversion of the sign of rotation between 360 and 340 nm and the big increase in the size of rotation at shorter wavelengths.

EXAMPLE 3:3. Activity of $(\pm) = \frac{1}{1.77} \times$ activity of $(-)$; activity of $(+) - 0.130 \times$ activity of $(-)$; stereospecific index $= 7.7:1$.

If the activity of $(+) = \frac{1}{4.6} \times$ activity of $(-)$, activity of $(\pm) = 0.609 \times$ activity of $(-)$ and equipotent molar ratio is 1.64. The error in the assay is 8%.

EXAMPLE 3:4. From the rotations

$$p = \frac{7.7+11.5}{2\times 11.5} = 0.835 \quad \text{or} \quad \frac{143+118}{2\times 143} = 0.913.$$

The latter is more likely to be correct because the angles measured are bigger. The estimates of the activity of the pure $(+)$-isomer are 15.2 and 13.9. If the weaker base were incompletely resolved and x is the activity of the stronger base, the observed ratio

$$R = \frac{x}{(1-p)x+p} \quad \text{and} \quad x = \frac{Rp}{1-(1-p)R}.$$

At the two wavelengths the estimates of p and $1-p$ are, for $R = 5.7$,

$$x = \frac{5.7(0.8348)}{1-5.7(0.1652)} = 81.5 \quad \text{and} \quad \frac{5.7(0.9126)}{1-5.7(0.0874)} = 10.4.$$

For $R = 6.7$,

$$x = \frac{6.7(0.8348)}{1-6.7(0.1652)}$$

which is negative, but

$$\frac{6.7(0.9126)}{1-6.7(0.0874)} = 14.7.$$

For $R = 11.0$ the higher estimate of stereochemical purity gives $x = 260$.

EXAMPLE 3:5. From the rotations

$$p = \frac{1105+1184}{2\times 1184} = 0.9666.$$

If K for pure $(+)$-form $= x$, $0.967x + 0.0334(5.035 \times 10^9) = 4.188 \times 10^8$ and $x = 2.506 \times 10^8$ ($\log x = 8.399$). The 'corrected' stereospecific index $= 20.1$ and the limit to stereochemical purity

$$= \frac{20.1}{21.1} = 0.953.$$

EXAMPLE 3:6. 4; the methyl ester at one end destroys the plane of symmetry.

EXAMPLE 3:7. With reference to glyceraldehyde (Ar → CHO) it is L but with reference to serine (OH → NH$_2$ and CH$_2$NHMe → COOH) it is D. The description R is unambiguous.

EXAMPLE 3:8. Both are S.

EXAMPLE 3:9. 2S, 3S, 4S.

EXAMPLE 3:10. There are 4 asymmetric carbon atoms but those in positions 1 and 5 are locked by the pyrrolidine ring so there are only $2^3 = 8$ isomers, the enantiomeric pairs of ecgonine (hydroxyl group equatorial, carboxyl group axial), *pseudo*ecgonine (hydroxyl group equatorial, carboxyl group equatorial), *allo*ecgonine (hydroxyl group axial, carboxyl group axial), and *allopseudo*ecgonine (hydroxyl group axial, carboxyl group equatorial). $(-)$-Cocaine is 2R-methoxycarbonyl, 3S-benzoyltropine. It is actually a derivative of *pseudo*tropine but this is indicated by the 3S arrangement. The locked positions at 1 and 5 are sometimes written in 1r, 5r and the enantiomers will have a 1s, 5s arrangement.

EXAMPLE 3:11. 3S, 5R, 6R.

EXAMPLE 3:12. 4; 4. PGE$_2$ is 5Z, 8R, 11R, 12S, 13E, 15S. PGF$_{1\alpha}$ is 8R, 9S, 11R, 12S, 13E, 15S.

EXAMPLE 3:13. Cortisone, 6; C8, C9, C10, C13, C14, C17. Aldosterone, 7; C8, C9, C10, C11, C13, C14, C17.

Chapter 4

EXAMPLE 4:2. The graph of logit(y) against log P bends at each end. The middle range has a slope of 1.8 and when logit$(y) = 0$, $\log P = 1.21$ (approx.) so $\log K = -(1.8 \times 1.21) = -2.18$ and $K = 0.0066$ (mm Hg^{-1}): n is not constant.

EXAMPLE 4:3. The % inhibition is:

1 nm	3 nm	10 nm	20 nm	30 nm	100 nm	300 nm	1 μM
14.9	23.2	49.2	65.5	72.2	88.4	95.6	99.1

With θ_1 less than 0.01 and 49% inhibition at 10 nM, $\log K \doteqdot 8$. From the Hill plot $\log K = 7.98$; the values at either end lie off a straight line but there are 5 points which form a good straight line with a slope of 0.96. Values have not been corrected for depletion of the ligand when there is high uptake (with low concentration of inhibitor). Depletion of the inhibitor is negligible in this experiment.

EXAMPLE 4:4. The graphs of logit(% bound or % inhibition) against log dose are better straight lines than the graphs of % bound or % inhibition against log dose. The effects produced by the standard and the homogenates indicate that approximately one gland contains 900×108 ng LH but the homogenates produce more inhibition of binding than the corresponding standards so the glands actually contain slightly more than this. At 50% inhibition the increment in log dose which would produce a match is 0.18 to 0.20, depending how the results are plotted, and the estimate is 147 to 154 µg/gland. A least-squares fit to the Hill plots gives an estimate of 153 µg/gland from the intercepts on the x-axis. Treated as a bioassay the results are 140 µg/gland with the larger dose and 147 µg/gland with the smaller ones. The difference is because the lines for standard and test do not appear to be parallel, possibly there are other substances present in the gland which bind to the protein to some extent and interfere with the assay. Note also that the slopes of the Hill plots are greater than 1 (1.39 for standard and 1.21 for homogenate).

EXAMPLE 4:6.

Concentration		Both substrates together					Tyramine alone	
Tyramine	Tryptamine	θ'_1	V'_1	θ'_2	V'_2	Total	θ_1	V_1
1 mM	1 mM	0.349	12.0	0.319	7.4	19.4	0.513	17.6
2	2	0.419	14.4	0.383	8.9	23.3	0.678	23.3
5	5	0.475	16.3	0.434	10.1	26.4	0.840	28.8
10	10	0.498	17.1	0.455	10.6	27.7	0.913	31.3

The ratio θ_1/θ'_1 increases in the order 1.47, 1.62, 1.76, 1.83, i.e. the inhibitory effect of the tryptamine increases with concentration and above 2 mM this is obvious because the rate is less than with tyramine alone.

EXAMPLE 4:7. The slopes are 1.015 for the controls, 1.050 in the lower concentration of inhibitor and 1.002 in the higher concentration.

EXAMPLE 4:8. At 10^{-9} M, $DR = 4.4$; at 10^{-7} M, $DR = 341$.

EXAMPLE 4:9. The stereochemical purity is at least $\frac{1000}{1001} = 99.9\%$. The dose-ratios and apparent affinity constants are:

			Alone		Together	
Conc.	Obs DR	K^*	DR_w	DR_s	Comp.	Non-comp.
10^{-5}	25.0	2.40×10^6	6	26	31	156
5×10^{-6}	24.9	4.78	3.5	26	28.5	91
10^{-6}	16.5	1.55×10^7	1.5	25	25.5	37.5
5.5×10^{-7}	24.6	4.29	1.3	26	26.3	34.2
2.5×10^{-7}	22.4	8.56	1.1	22	22.1	24.2
1.2×10^{-7}	28.1	2.26×10^8	1.06	31	31.1	32.9

The graph is quite a good straight line and the compounds appear to behave competitively, though with more than about 10% of the stronger enantiomer present the combined dose-ratios for competitive and non-competitive behaviour are not very different.

EXAMPLE 4:10. The matching concentrations are 0.994×10^{-5} M (A_1), 3.064×10^{-5} M (A_2), so $1-p = (3.064-0.994)/3.2 = 0.647$; $p/(1-p) = 0.546$ and $K_p = 0.546/5 \times 10^4 = 1.09 \times 10^3 \,\mathrm{l\,mol^{-1}}$.

EXAMPLE 4:11. The matching concentrations are:

P	A
3×10^{-3} M	1.41×10^{-5} M
6	2.42
12	3.43
24	4.16

A least-squares fit with a double-reciprocal plot gives $K = 99$; with direct fit to a hyperbola $K = 121$. The concentration of P which should produce a response of 26 is 20.9×10^{-3} M so

$$p = \frac{PK_p}{1+PK_p} = 0.674 \quad (\text{or } 0.717)$$

and the efficacy is 1.48 (or 1.40). The efficacy must be greater than 1 if the compound can produce a response greater than half the maximum of which the tissue is capable.

EXAMPLE 4:12. The dose-ratios give the following values:

	Offset	Onset
Time	y_t	y_t
0	0.85	
1	0.78	0.22
2	0.69	0.40
3	0.63	0.57
4	0.58	0.67
5	0.53	0.73
6	0.50	0.78
7	0.48	0.81
8	0.42	
9	0.39	
10	0.36	$y_e = 0.85$

A least-squares fit of $\ln y_t$ for offset against t gives $k_{-1} = 8.44 \times 10^{-2}\,\mathrm{min}^{-1}$ and a least-squares fit of $\ln(y_e - y_t)$ against t for onset gives

$$k^* = 4.58 \times 10^{-1}\,\mathrm{min}^{-1} = k_{+1}A + k_{-1},$$

so

$$k_{+1} = \frac{0.458 - 0.0844}{1.5 \times 10^{-9}} = 2.49 \times 10^8\,\mathrm{min}^{-1}\mathrm{l}\,\mathrm{mol}^{-1}.$$

$$K = \frac{k_{+1}}{k_{-1}} = 2.95 \times 10^9;$$

from equilibrium dose-ratio

$$K = \frac{5.67}{1.5} \times 10^9 = 3.78 \times 10^9\,\mathrm{l}\,\mathrm{mol}^{-1}.$$

EXAMPLE 4:13. $8 \times 10^{-8}\,\mathrm{M}$; dose-ratio $= 83$; receptor occupancy $= 98.8\,\%$.

EXAMPLE 4:14. $7 \times 10^{-7}\,\mathrm{M}$; dose-ratio $= 3.1$; receptor occupancy $= 67.7\,\%$. The high dose-ratio with atropine ensures protection of the heart from vagal arrest as well as prevention of excessive bronchial secretion and is not itself dangerous. With $(+)$-tubocurarine there is a need for the minimum relaxation necessary to help the surgeon; even this impairs respiration. The concentration of ethyl alcohol equivalent to 80 mg% $= 1.7 \times 10^{-2}$ M, which is more than 200,000 times the concentration of atropine used. It is acting by a physicochemical mechanism rather than by one involving receptors.

Chapter 5

EXAMPLE 5:1. The graph of $\log N$ against $\log P$ is a good straight line, with the equation $\log N = 2.098 - 0.949 \log P$ and no point off it by more than 0.2 log units; $r = -0.988$. The graph of $\log N$ against $\log p_s$ suggests that there is a trend. The very active compounds lie below the line and the observed value for chloroform, for example, indicates twice as much potency as corresponds to the calculated value; $r = 0.955$. The graph of $\log P$ against $\log p_s$ has $r = -0.964$ so the biological results appear to fit better than the chemical results!

EXAMPLE 5:2. $p_t/p_s = 0.004(760)/104 = 0.029$: $p_t = \frac{243}{104}(0.4)\%$ $= 0.93\%$.

EXAMPLE 5:3. If the results are in a geometrical progression,

$$C_{hex} = (C_{pent} \times C_{hept})^{1/2} = (0.007 \times 0.00038)^{1/2} = 0.0016 \text{ M}.$$

If a least-squares fit is made of log concentration against chain length, n, the equation is $\log C = 0.839 - 0.598n$, so the value for n-hexanol is -2.749 and $C = 0.0018$ M. The correlation coefficient, $r = -0.999$ but this applied only to the range n-propanol to n-octanol (see Figure 5.1A).

EXAMPLE 5:4. The solubility appears to be divided by almost exactly 4 for each extra methylene group so the concentration would be 0.001 M for nonanol and 0.00025 M for decanol. The concentration of hexanol inhibiting *Staph. aureus* would be $\frac{0.25}{0.8} \times 0.25 = 0.078$ M (which is greater than the solubility). Each methylene group alters the concentration inhibiting *S. typhosus* by about 0.29 so nonanol would be active in a concentration of about $0.0034 \times 0.29 = 0.00099$ M and decanol in a concentration of about 0.00029 M (which is greater than the solubility). A least-squares fit of $\log C$ against n gives: $\log C = 1.761 - 0.528n$, so the concentration for nonanol would be 0.0010 M and for decanol it would be 0.00030 M. The value of $r = -0.999$.

EXAMPLE 5:5. For the steps

$$n\text{-pent}\overset{+}{\text{N}}\text{Me}_3 \rightarrow CH_3COOCH_2CH_2\overset{+}{\text{N}}Me_3 \rightarrow PhCH_2COO—$$
$$\rightarrow Ph(C_6H_{11})CHCOO— \rightarrow Ph(C_6H_{11})C(OH)COO—$$

The upper limit to $\log K$

$$= 3.7 + 0.7 + 3.5 + 4.0 + 1.9 = 13.8;$$

the lower limit to $\log K$

$$= 3.7 - 0.6 + 0.6 + 0.9 - 1.5 = 3.1,$$
$$\phi_v^0 = 189.7 - 13.8 + 63 + 79 + 1 = 318.9$$

Index

forces between drug and receptor 181–5, 218–20
free energy, *see* Gibbs free energy
Freundlich isotherm 117–18
full agonists 154

Gaddum 1, 157–8, 187
Gaddum-Schild equation 158, 162, 180
gauche (conformation) 107–9, 113–14
Gibbs free energy 41–2, 189
    and equilibrium constant 42
    and Hammett relation 77
    increments in 77, 191, 203
(+)-glyceraldehyde 93

half-life 50
haloethylamines 181
Hammett relation 70–4, 77–9
    free energy and 77–9
    predictive use of 74
Hansch 75, 196–203
    equation 197
Hill plot 121, 140, 149, 176–7
Hofstee plot 139
homologous series
    of alcohols 191–4
    of trimethylammonium salts 153–4, 215–16
    physical properties in 191–4
hydrogen bonds 183, 214, 218
hydrophobic bonding 184, 213, 220
hydrophobic effects 183

I_{50} 145
index of discrimination 30
index of significance of slope 18
infrared spectra 104
inhalation, limits to 53, 188
inhibitor constant 144
    measurement of 144–9
interquartile range 10, 13
ionic product of water 65
ionisation 64–7
    and biological activity 67–8
    and drug transport 67–8
isomers 85
    geometrical 101
    optical 85
    purity and activity of 88–91, 105

Karplus relation 113
kidney, excretion and the 61–2, 67

K_m (Michaelis constant) 138
    apparent 148
        competitive inhibitor and 148
        noncompetitive inhibitor and 148
    measurement of 138–140

Langmuir isotherm 118
    linear transformation of 119–21
Latin square 21
Law of Mass Action 39, 151
LD_{50} 24–7
least-squares fit *see* direct fit
    to Hansch type polynomial 227–8
    to hyperbola 228–30
    to logistic expression 230–1
    to straight line 224–7
linear transformation of hyperbola 119–21
    errors attached to 138–40
Lineweaver-Burk plots 138–9
    with inhibitor present 147
log dose-response curve 15
    slope of 30, 153
logit 120, 140
log normal distribution 24, 26

Mann-Whitney (Wilcoxon) statistic (U) 11–12, 38
maximum effects of groups (on affinity) 210–15
maximum response (of tissue) 153–4
    receptor occupancy and 151, 154
    noncompetitive antagonists and 161–2
mean 4
median 10
membranes, transport across 60–2
metaphilic effect 175
Michaelis constant 138, 148
    measurement of 139–40
molality 39, 81
    and molarity 82
molal volume (V_m) 80, 82, 201
    apparent (ϕ_V) 81–2, 207, 211–12
molar activity (biological) 30
molarity 39
mole fraction 39
molecular models 83
molecular orbitals 109–11

narcosis 188, 193
Newman projection formulae 107–8
nitrogen mustards 182